Total Solar Eclipse 2017

Your Guide to the Next US Eclipse

Marc Nussbaum

Copyright

Warning and Disclaimer

This book is designed to provide information about eclipses in general and the August 21, 2017 Eclipse in particular for people who may want to observe the event in person. Every effort has been made to make this book as complete and as accurate as possible, but **no warranty of fitness is implied**. Viewing the Sun without proper eye protection can result in severe eye damage including blindness. The Author and Publisher of this book provide information and advice with regard to proper eye safety and eclipse viewing procedures, however neither party shall be held responsible for the accuracy and efficacy of the eye safety advice or related information contained herein, or omissions with regard to eye safety. It is the responsibility of the reader to ensure proper precautions are taken when viewing the eclipse. This information is provided on an as-is basis. The Author and Publisher shall have neither liability nor responsibility to any person or entity with respect to any loss or damages arising from the information contained in this book or from the use of the websites or products that are suggested within.

About the Author

Marc Nussbaum

Marc Nussbaum is an author and lecturer on science topics including eclipses, astronomy and space exploration. He is Founder, President and CEO of Audible Rush, Inc. a manufacturer of speaker systems for outdoor recreation and sporting applications. Prior to launching Audible Rush, he served as CEO of Lantronix Corporation, a publicly traded NASDAQ manufacturer of networking equipment for connecting machines to the Internet. He was also co-founder of Western Digital's hard drive business, serving as the company's lead technology executive in the positions of Senior Vice President Engineering and Chief Technical Officer.

Marc chairs the science curriculum committee for the University of California Irvine Extension, Osher Lifelong Learning Institute (OLLI). He is a photographer, computer engineer, and an amateur astronomer. Marc holds a BS degree in Physics from the State University of New York.

This book was written to encourage as many people as possible to go out and observe a total solar eclipse for themselves. Marc and his wife Sherri live in Irvine California.

To reach Marc by email: marcnussbaum@audiblerush.com.

Acknowledgments

I thank my amazing wife and business partner Sherri for her advice, patience, encouragement and most of all her smiles. I've been incredibly fortunate to be the recipient of her "I love you smile" dozens of times a day for 34 years—every single time, I light up inside. She's known around town as "the make it happen chick" and without her our business would not exist, Orange County, California would be a lot less fun place to live, and thousands of people would have missed out on her generous gifts of time and caring. I love you so much. Thank you for your advice, your editing, and all your smiles. Everything is awesome; we're living the dream.

Secondly, a big thanks to my father, Bob Nussbaum, who taught me about the importance of thinking independently and not being afraid to be different. As an entrepreneur, Dad launched several retail stores and worked as a skilled restoration craftsman in his own business as the Orange County Furniture Doctor. Today at 86 years of age, he is an invaluable sounding board for my off-the-wall ideas and has made numerous contributions to this book and other projects.

Next is a thank you to my son Eric who has supported me by being the first person to "get" my ideas. Thanks for always finding the time to help out. Also thank you for letting me teach you a few things along the way. I often think back to those countless special moments: afternoons at the big pool, steadying your bicycle, patiently learning the motorcycle clutch, trips down the ski slope and days rafting down the river.

A special thank you to my big sister Elisa Delman whose encouragement and love is ever present, even with three thousand miles between us. Elisa and her husband at the time first introduced Sherri and me to eclipse chasing. She convinced us to join her on a Caribbean cruise to watch the total solar eclipse of February 26, 1998. Thank you, Sis.

Along with Sherri, Dad, Eric and Elisa, several friends graciously contributed their time and expertise to this book, providing invaluable comments and corrections to the manuscript. A special thank you goes out to Craig Small who continues to inspire, having missed only a handful of totality events over the past 50 years. His eclipse chasing adventures are legendary. Also thank you to Teri Bellows, David Chester, Janet Dreyer, David Fischer, Bill Frank, Marc Goldstone and Rick Hull for your advice, encouragement and most of all for generously contributing your valuable time and expertise to this project.

Key details in this book depend upon the models and analyses of a small handful of professionals working in astronomy and meteorology. Eclipse predictions and eclipse track map coordinates in this book are courtesy of Fred Espenak, NASA/Goddard Space Flight Center (http://eclipse.gsfc.nasa.gov/eclipse.html). Also thank you to Jay Anderson for his cloud and weather data (http://www.eclipser.ca).

Trademarks

All terms mentioned in this book that are known to be trademarks or service marks are appropriately capitalized. The author cannot attest to the accuracy of this information. Use of a term in the book should not be regarded as affecting the validity of any trademark or service mark.

Lightroom™ and Photoshop™ are trademarks of Adobe Inc. AstroSolar™ is a trademark of Baader Planetarium Gmbh. Gorillpod™ is a trademark of Joby. PowerShot™ and EOS™ are a trademark of Canon. Lumix™ is a trademark of Panasonic. Stylus™ is a trademark of Olympus. Coolpix™ is a trademark of Nikon. Alpha™ is a trademark of Sony. iPhone™ is a trademark of Apple. Android™ and Google Earth™ are trademarks of Google. SnapZoom™ is a trademark of Snapzoom. HookUpz™ is a trademark of Carson Optical.

Table of Contents

Dear reader,

I am thrilled that you have selected "Total Solar Eclipse 2017" as your guide to this not-to-be-missed experience!

After completing the book, I would be forever grateful if you would take a moment and post a brief review on Amazon. These reviews not only help determine a book's success, they also are a great way to provide feedback and suggestions for future releases.

If you are not sure what to write, just leave a few sentences about what you liked and what you would like to see improved. Every comment is helpful and potential readers will appreciate your insights.

My goal is to encourage as many people as possible to experience the 2017 total solar eclipse. Thank you for taking the time to help me spread the word.

To leave a review, enter the following link in your web browser: http://amazon.com/author/marcnussbaum

From there, click on the book cover which will take you to Amazon's homepage for the book. Scroll down past the product details section, and click on the "Write a customer review" box.

Thank you,

Marc

Introduction

On Monday, August 21, 2017, the universe will reveal itself in an epic show more astonishing than anything ever devised by the magic of Hollywood or Disney. This event will surpass all the special effects we've ever created—simply because it will be real. Mother Nature's multi-gigaton, 24/7 nuclear fusion inferno in the sky, is going to put on a show.

Sure, you've seen an eclipse before, but you have never witnessed anything like this. You are about to see a TOTAL solar eclipse. The Sun gives rise to all life on Earth, producing enough energy every second to power New York City for a hundred years.[1] On eclipse day, when the Moon and Sun conspire to play hide-and-seek, we here on Earth will see what the Sun has been concealing all these years. As the blinding light goes dim, we get to peek behind the wizard's curtain and see for ourselves, in person—the solar atmosphere.

Unlike a rock concert, the tickets to this show are free! You don't have to be a member of the one percent to see it. There is no gatekeeper to the event. No bouncer will assess your sense of style and pass judgment. No radio station to call, in hopes of winning giveaway tickets. The best seats in the house cost the same as in the nose bleed section. You can even bring your own drinks (and a picnic lunch). There are no corkage fees either.

Front row is located right here in the U.S.A. The shadow will sweep through 12 states while making its way from the Pacific to the Atlantic Ocean. If you know where to sit, the show will be spectacular. This book will guide you to the hottest seats with the most promising weather prospects. We will also make sure you understand what is happening backstage, so you can fully appreciate the magic and science behind the beauty.

This book is for those of you who have heard about the upcoming 2017 Eclipse and are wondering what all the fuss is about. Explanations are kept simple. All information is scientifically accurate and will serve to build a strong base for further inquiry.

Even if you have little or no science background, by the end of the read you will be comfortable with the astronomical terms used by amateur and professional eclipse chasers. You'll be throwing around words such as *umbra*, *solar prominences* and *Baily beads*. What's more, you will know what they mean and when to look for them during the eclipse. The first few times words and phrases are used they are placed in italics to bring them to your attention; for example, *umbra*.

If you have a science background but are a few years removed from your last astronomy class, this book should serve as a fun reminder of all you've forgotten about the orbits of the Earth, Moon and Sun.

- Chapters 1 through 5 explain what a total solar eclipse is all about and how eclipses appear to us here on Earth.

- Chapters 6, 7 and 8 contain the information necessary to plan and pack for your eclipse adventure and instructions on how to view it safely. We help you pick the best observation site and provide a list of what to bring.

- Chapters 9 through 11 extend your knowledge beyond the basics. We cover a brief history of the eclipse (Chapter 9), answer advanced questions about the orbital motion of the Moon, Sun and Earth (Chapter 10) and provide guidance on photographing the eclipse (Chapter 11).

- Chapter 12, "Resources," contains a list of our favorite websites for planning and purchasing specialized gear such as eye protection and camera filters. Links are provided for local maps, eclipse weather forecasts, organized trips and more.

After reading this book, you will have the necessary knowledge to plan an amazing trip to your VIP seats on August 21, 2017.

Happy observing.

Chapter 1. It's Amazing Out There:
What is a Total Solar Eclipse and Why Bother?

You may be new to eclipses. More likely, you are new to this thing called a *total solar eclipse*. It's easy to forget high school astronomy, so you are bound to have plenty of questions. Before we dive in for a closer look, let's answer some of the big ones right up front.

What Is It Like?

During a total solar eclipse, the Moon lines up in front of the Sun and, for a short time, covers it completely. Day turns into night, and you can look directly towards the Sun without being blinded by the brightness. For the first time, you see the solar atmosphere—the fantastic *corona* spreading out far into space like the petals of a cosmic flower. Most people agree it is the most beautiful astronomical event we can experience here on Earth.

Don't These Things Happen All the Time?

There is a total solar eclipse on Earth about once every 16 months;[2] however the majority of them touch down in inaccessible or politically unfriendly places across the globe. On average, one occurs in the mainland United States only three or four times in our lifetimes. The last one occurred almost 38 years ago. After the 2017 Eclipse we'll have to wait until 2024 for the next one; after that, we'll have to wait another 21 years, until 2045.

When and Where Do I Need to Travel?

The next U.S. total solar eclipse will be August 21, 2017 (a Monday). Unless you happen to be one of the lucky few who live directly under the narrow funnel of the *shadow path* as it sweeps across the country, you will need to make travel plans.

This *shadow of totality* will pass through portions of Oregon, Idaho, Wyoming, Nebraska, Kansas, Missouri, Illinois, Kentucky, Tennessee, Georgia, North Carolina and South Carolina. In Chapter 6, "Planning the Eclipse" we will help you select the best place to watch the event unfold on eclipse day.

"The city I live in is not in the direct path of the eclipse shadow, but we are close. Can I watch it from home?" No. To fully experience the event you need to get yourself under the shadow into the *path of totality*. Step ten feet outside the edge of the shadow and you might as well be 2,000 miles away. You will miss all that is unique and spectacular about the event.

If you are unable to travel, I'm not saying you shouldn't venture outside your house and view a *partial* version of this eclipse. However, the Sun won't be completely covered, so don't expect to be wowed.

By the way, you need to use proper eye protection during any solar eclipse. Be sure to read Chapter 8, "Safely Viewing the Eclipse."

How Long Will the Eclipse Last?

Total solar eclipses last about three hours but the majesty of totality, when the Moon completely covers the Sun, only lasts for a few minutes. Several factors determine the length of totality, the major one being how large the Moon appears in the sky. The larger the Moon, the longer it will fully cover the Sun. Of course the size of the Moon does not change, but because of its elliptical orbit it is sometimes closer to the earth and during that time it appears larger.

Duration of totality also varies depending where along the shadow path you are located. Under the centerline traced by the shadow, totality for the August 21, 2017 Eclipse will vary from 2 minutes, 0 seconds to 2 minutes, 40 seconds.

Can I Bring the Kids?

Absolutely yes. Pull them out of school or summer camp and take them. This eclipse will be an incredible experience. Even today's kids, jaded by all their electronic games and action movies will be amazed. It's like watching the moon landing back in 1969 or seeing a space shuttle launch in person; they'll likely remember it forever.

If your child does remain in camp or school during the eclipse, he or she may miss the experience of a lifetime for several reasons. It is unlikely the camp or school would transport the children to the shadow path of the eclipse, the only area where totality will be visible. Furthermore, instructors and teachers who are unsure of safe-viewing procedures are concerned about liability and will not allow children to even attempt to view the eclipse.

If you're lucky, the school or camp will set up a pinhole observation arrangement so the class can watch the Sun's projection on a piece of cardboard (or inside a shoebox). Viewing a projection works fine for when the Sun is partially covered, but it doesn't work at all when the Sun is covered completely during a total eclipse. This means they miss seeing the Sun's phenomenal corona and other amazing sights.

For these reasons, I urge you to take your children or grandchildren to witness first-hand their first total solar eclipse. Don't miss Chapter 8, "Safely Viewing the Eclipse," and don't forget to bring the camera and capture their looks of amazement as totality begins. It's like visiting Disneyland for the first time.

In addition to learning some astronomy, youngsters will come away understanding that they can explore, first-hand, the world of science and nature. Who knows? You might even spark a lifelong love for scientific inquiry.

Why Should I Make the Effort to See One?

Thousands of people from all over the world will be traveling to see this event. Witnesses to totality will talk about it for years to come. For the rest of your life, there will be two groups of people: those that saw totality in 2017, and those who don't understand why they can't even join the conversation. Think about it. There are people that attended Woodstock and people that didn't.

True story: My wife and I flew to Australia and boarded a cruise ship scheduled to observe the November 13, 2012 total solar eclipse. Totality started at 5:45 AM and lasted about 2 minutes. Light clouds parted as the shadow approached. It was magnificent.

The night before, we shared our dinner table with an Australian couple who had chosen the cruise for reasons other than the eclipse. I explained the rarity and magnificence they were about to see. The next night I asked what they thought of the experience. The dear woman explained that it was too early in the morning, so they didn't bother to get up to see it. We found new table companions for the rest of the cruise.

Don't miss the rare opportunity to witness nature's most spectacular astronomical spectacle in 2017!

Reason #1 to View the Total Solar Eclipse: Epic Emotion and Beauty

A total solar eclipse is an incredibly moving and otherworldly experience. It's Mother Nature flexing some big muscles. Have you ever been at a live concert, musical or opera and felt a deep emotional reaction as a particular singer's voice reached out and grabbed you? Now that's what I'm talking about. Something about witnessing totality instantly connects you to mankind's evolution and struggle through history. Without being consciously aware, your mind conjures up dinosaurs walking the earth, cavemen around a fire, the pyramids, biblical people roaming the Sinai desert or the interior of a medieval cathedral. Don't ask me why, it just happens. While I can't guarantee this will be your experience, I am willing to bet you will feel emotions and think of things that will surprise and delight you. You might even cry a little. Seeing one is deeply moving.

Totality is the stuff of science fiction. This is your best chance to get a taste of what it is like to stand on the bridge of a real "Starship Enterprise." I hope "The Great Bird of the Galaxy" (Star Trek creator Gene Roddenberry),[3] and the rest of the Star Trek cast and crew made time to observe totality at least once in their lives.

Reason #2 to View the Total Solar Eclipse: You Have Never Seen One Before

There are five different types of eclipses. We will discuss how each of them are formed in Chapter 4, "Up There: How eclipses happen."

Of all the types, only a total solar eclipse blocks out the Sun with a perfect-fitting Moon, revealing the Sun's corona. The unique, spectacular nature of totality when the Sun's atmosphere is revealed is hard to forget. It stands above any other astronomical event we humans will witness in our lives.

We have all seen eclipses before, but I'm going to go out on a limb and bet that you have never seen a total solar eclipse. You might think you have, because television and newspapers have confused us all.

Since a total solar eclipse is the only type that reveals the Sun's prominences and spectacular corona, I wish we called them something that made this more obvious. As we'll discuss in detail later, an eclipse can be a partial lunar eclipse, a total lunar eclipse, a partial solar eclipse, an annular solar eclipse or finally, a total solar eclipse.

A *partial* solar eclipse is like a total solar eclipse the same way a hole in the ground is like the Grand Canyon. Sure, they have things in common—perhaps they were both created by the same phenomenon of river erosion. However, you would miss out on a very cool experience by assuming the Grand Canyon is just a bigger version of the hole in the ground.

As a child, you might remember making one of those shoebox projectors I mentioned earlier as a school project. The eclipse that passed through your neighborhood (or school yard) that day was almost certainly not a total eclipse. Perhaps you helped your child with one of these experiments. While it was educational, you may have thought it was a bit underwhelming. Don't let an earlier observation experience of watching a dull partial eclipse dissuade you from getting out to see *totality*.

Reason #3 to View the Total Solar Eclipse: Perhaps Your "Last Chance to See" (In the United States)

Douglas Adams, author of the irreverent "The Hitchhikers Guide to The Galaxy," also wrote and presented a non-fiction BBC radio show titled "Last Chance to See."[4] The series chronicles his adventures traveling to far off places in an attempt to encounter various animal species that are soon to be extinct. He visits the manatees in the Amazon, white rhinos in the Congo, mountain gorillas in Uganda and Komodo dragons in Indonesia to name a few.

It's not that total eclipses will become extinct in our immediate future, but as individuals we have a small number of chances to see one in our limited lifetimes.

In the original edition of "Last Chance to See," Adams asks Australian venomous reptile expert Dr. Struan Sutherland, whether there are any venomous creatures that he likes. The expert replies, "There was, but she left me."[5]

On average, there is a total eclipse every 16 months; however these events are spread out all over the globe. Of the last 20 solar eclipses, only 3 were total eclipses that fell in the contiguous United States. The last time we had a total in the country was 38 years ago (February 26, 1979). The main shadow brushed a few states in the Northwest US and unfortunately weather conditions were disappointing for most observers.

In 2017, totality will be visible in our collective backyard as it sweeps across the entire US heartland. Even if you live as far south as Houston, or as far north as Seattle, the *path of totality* is just a day or two drive on well-maintained US highways. The last time a total eclipse crossed such a large portion of the country was 99 years ago (June 8, 1918).

Looking ahead to the end of the century, over the next 84 years there will be only seven total solar eclipses in the contiguous United States:

1. August 21, 2017 (12 state coverage: Oregon, Idaho, Wyoming, Nebraska, Kansas, Missouri, Illinois, Kentucky, Tennessee, Georgia, North Carolina and South Carolina)

2. April 8, 2024 (8 state coverage: Texas, Arkansas, Missouri, Indiana, Ohio, New York, Vermont and Maine)

3. August 12, 2045 (10 state coverage: California, Nevada, Utah, Colorado, Kansas, Oklahoma, Arkansas, Mississippi, Alabama and Florida)

4. March 30, 2052 (2 state coverage: Florida Panhandle and Georgia)

5. May 11, 2078 (5 state coverage: Louisiana, Alabama, Georgia, North Carolina and South Carolina)

6. May 1, 2079 (4 state coverage: Pennsylvania, New York, Rhode Island and Massachusetts)

7. September 14, 2099 (9 state coverage: North Dakota, Minnesota, Wisconsin, Michigan, Indiana, Ohio, West Virginia, Virginia and North Carolina)

Keep in mind, that depending on where you live, high demand may make travel by air difficult for the more geographically concentrated eclipses. Another thing to consider is the likely weather outlook during the month of the event. August is better than April. It only takes one storm in your part of the country to destroy your chance to see the eclipse.

Also, over the 84 years covered in the list above, (2015 to 2099), only two total eclipses will be visible in the states north and west of Texas (2017 and 2045). If you live in this part of the country and miss the 2017 Eclipse, you'll have to wait 32 years for another chance in 2045. Looking past the year 2045, the next total solar eclipse to sweep through the western US is 65 years later in 2110!

Using the most recently available Social Security Actuarial Life Table, if you were more than 82 years old in 2015, the total solar eclipse in 2017 may be your last chance to see one. Here's how many opportunities you have to see a total solar eclipse in the United States in your remaining lifetime (based on your age in 2015).[6]

- 82 year-old: 1 remaining eclipse (2017)
- 54-81 year-old: 2 remaining eclipses (2017 and 2024)
- 46-53 year-old: 3 remaining eclipses (2017, 2024 and 2045)
- 19-45 year-old: 4 remaining eclipses (2017, 2024, 2045 and 2052)
- 18 year-old: 5 remaining eclipses (2017, 2024, 2045, 2052 and 2078)

On a personal note, as I write this, I am 59 years old and live in Southern California. Based on these Social Security tables, the 2017 Eclipse is my "last chance to see" totality without traveling outside the western United States.

Eclipse trivia: The August 21, 2017 eclipse will be the first total solar eclipse to make land ONLY in the US (no foreign territories). That makes it the world's first "All American Eclipse." The last one that had a similar landfall profile was 760 years ago, but since the United States did not exist back in 1257, that makes this the first ever "All American." The next "All American" won't occur until January 25, 2316.

Reason #4 to View the Total Solar Eclipse: It's an Adventure

We all have a collection of unique personal experiences that make us colorful and interesting individuals. A total solar eclipse is a once in a lifetime grand adventure that you will never forget. Once experienced, it will forever be a source of "internal smiles." What adventures are still on your bucket list?

Flying airplanes had been a dream of mine ever since I could remember. As a 17-year-old high school student, a buddy of mine and I piloted a four seat, single engine prop plane 3,000 miles round-trip to watch the last Skylab launch from Cape Kennedy, Florida. We almost ran out of fuel over Maryland and ended up with a military fighter escort after an accidental detour into restricted airspace. We landed at Titusville Airport, hitched a ride to the viewing causeway and witnessed a launch of the mighty Saturn-IB rocket. Our return trip was thankfully uneventful.

What's the point of this story? It was an adventure. Not quite in the same league as exploring the Arctic, competing in the Iditarod or climbing Mount Everest. But still, a great story for around the conference room table, at a cocktail party or to tell that 20 something at the adventurer's club over a Scotch served neat. Give yourself an adventure. See a total solar eclipse. I guarantee that you will have an interesting story to share with others.

Served "neat": A single unmixed liquor served unchilled, without water or ice. Alternative to served "on the rocks" (with ice). Alternative to served "up" (shaken with ice and then strained into a glass). Sometimes the term "straight up" is used to order "neat" but can be confused with "up." [7]

Can't I Just Watch It On TV?

Sorry, you cannot see one of these on TV. Well, you can, but you shouldn't. Imagine going back in time to 1969. Would you rather be watching the Moon landing on a small black and white TV screen or peering out of the Lunar Module window as Neil Armstrong took "one giant leap for mankind"? Yes, you can see it at home but, even your HD 1080P TV (or your 4K set) can't come close to transporting the experience into your living room. Someone wrote on a website recently that it would be like watching people eat a fantastic gourmet meal on television. You will have known that it happened, but the main point of the experience will have been lost.[8]

Watch it on TV and you will *totally* miss all the excitement (pun intended). Ask anyone on the planet who has seen totality and you'll get that same response in a nanosecond. I'm anxiously waiting for the first total eclipse program content to run on my virtual reality headset. By the way, I'm also anxiously waiting for an affordable virtually reality headset. I expect even our best technology will fall short of producing chills up my spine and tears in my eyes when the video-recorded corona lights up the sky.

Chasing Eclipses Internationally

If you are willing to travel overseas, opportunities for totality get much better. That's great news because after you've experienced the first total solar eclipse you'll want to see more. This is how we "carriers of the eclipse bug" infect other humans. Our mission is to recruit and indoctrinate new members into the unofficial global eclipse chaser society. There are varying levels of membership in this club. My wife and I enjoy combining international travel to places on our bucket list with an opportunity to see totality. So far, we've spent a week in the Caribbean, ten days in the Greek Islands and a month in Australia and New Zealand all timed to coincide with the occurrence of a total solar eclipse.

A few diehard members of the unofficial eclipse chaser society have been to every totality event on Earth since they were first bit by the bug. For some, the bug hooked them 50 years ago, back in college. With a total solar eclipse occurring somewhere every 16 months, there are plenty of opportunities for addicted *umbraphiles* to fully express their kink.

The large majority of eclipse chasers have daytime jobs to pay for their trips around the world. Indeed, it can become quite expensive to travel to some of these out-of-the-way places. Months before the eclipse airline flights can become over-booked and local hotels often approach capacity, driving up prices.

The dedication of these individuals is amazing. Eclipse chasers faced with the prospects of a particularly short totality or an inhospitable geography have occasionally banded together to charter a commercial jet. Flying in the same direction as the shadow extends totality in the same way driving a car in the same direction of a train prolongs the time the locomotive is in sight. On these flights, everyone gets a window seat on the side of the plane facing the eclipse. The rest of the seats usually remain empty.

If you start chasing early enough, it is possible to observe over 40 totalities in a lifetime. Some eclipse mavens also contribute a significant amount of their time and talent to help others experience the fun. A few publish eclipse bulletins, chair international astronomical organizations, provide eclipse weather guidance, maintain educational websites and sell the occasional guidebook or protective eye gear. I've provided a list of several of their online sites in Chapter 12, "Resources."

Total Solar Eclipses: 2001 - 2050

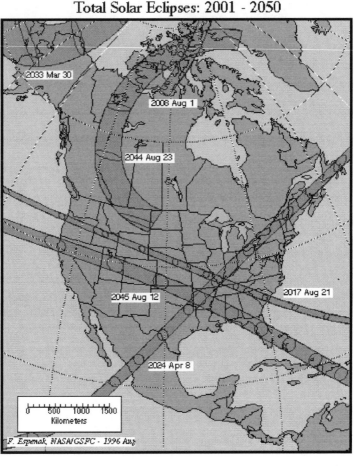

Upcoming total solar eclipses near North America.
Source: NASA/GSFC, Fred Espenak.

A riddle: Q- When is the Earth not the third rock from the Sun?

A- During a solar eclipse. Sun -> Mercury (1), Venus (2), The Moon (3), Earth (4), Mars, Jupiter, Saturn, Uranus, Neptune, Pluto (a "dwarf" planet).

This is also true while the Moon is closer to the Sun than Earth during half of the lunar orbit, when the Moon is between its third quarter and first quarter positions. See Chapter 10, "The Motions of the Sun, Earth and Moon" for more information about the lunar phases.

Chapter 2. The Big Day:
A Walk Through a Total Solar Eclipse

> Important Warning: Ensuring eye safety requires knowing this simple rule: ALWAYS USE PROPER EYE PROTECTION when looking at the Sun, UNLESS the Sun is COMPLETELY COVERED by the Moon.
>
> The only time you don't use eye protection (assuming you are in the path of totality), is DURING TOTALITY. Even if only a tiny portion of the Sun is still exposed, viewing it without eye protection may cause permanent eye damage. Proper eye protection can be very inexpensive, but you need to get it in advance of eclipse day. Please read Chapter 8, "Safely Viewing the Eclipse" to learn what constitutes PROPER eye protection. We also cover common mistakes people make when improvising eye protection at the last minute.

A Step by Step Guide to What You Will See

Viewing a total solar eclipse is a bit like going to a Broadway show. There are three distinct 'acts', lasting a total of about 3 hours. You put on your appropriate eye protection, the curtains part and...

Act 1. "Enter the Moon" (one long scene)

Act 1, Scene 1. The Company sings, "First Contact": The Moon and the Sun are on stage. They slowly work their way towards each other until—a first, somewhat tentative, kiss. The disks of the Sun and the Moon appear to touch for the first time; we have *first contact*. We watch as our two favorite actors in the sky start to embrace in a cosmic hug.

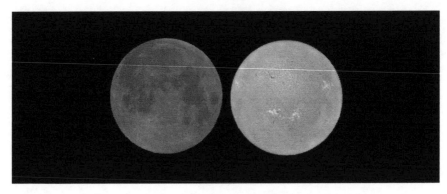

Just prior to first contact.

As things progress, we begin to realize that the Moon has another agenda. The kiss has turned into a bite! The Sun seems to be missing larger and larger pieces as it is eaten away by the Moon. This is sometimes referred to as the "Pac-man" phase. If you're not old enough to remember Pac-man, don't worry about it. The ancient Chinese thought a mighty dragon was devouring the Sun. In fact, during the first few minutes after *first contact*, astronomers say, "the Moon is taking its first bite out of the Sun."

You can hardly tell the Moon is moving, but each time you look up, a tiny bit more of the Sun is gone. Eventually, you notice half has been devoured. Surprisingly, the sky is still as bright as when this all started. As people around you chat quietly, the Moon continues to make a meal of the Sun.

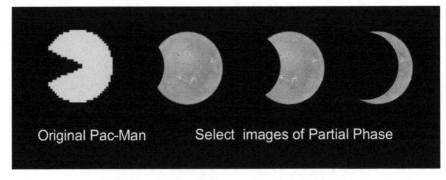

Original Pac-Man   Select images of Partial Phase

Partial phase progression of total solar eclipse,
with Pac-Man screen icon on far left.

After 30 minutes or so, only about a quarter of the Sun remains. You sense the tempo is picking up. While staring at the Moon for a few seconds straight, you can now clearly tell it is moving.

You notice the sky has changed; everything around you has become darker. People begin chatting excitedly. The Sun no longer lights the world stage. Colors have become grayer and muted, and then, something you've never seen before: a crescent Sun in the sky! It no longer looks familiar and instead appears more like a brightly lit object of artwork, an intense white sign someone hung up high. It doesn't look natural at all, and you think to yourself; "this thing is not supposed to be up there."

About 90% of the Sun is covered as the curtain goes down on Act 1.

Act 2. "Totality" (seven scenes that speed by too quickly)

In Act 2, fireworks fly, and the show reaches its climax. Although Act 2 will last less only a few minutes (about two-and-a-half minutes in 2017), it's what we've all been waiting for. Like a thrill ride, at the end we'll feel exhilarated and somewhat exhausted from the excitement.

Act 2, Scene 1. The Company sings, "Me and My Shadow": Totality is close, but the Sun isn't ready to disappear quite yet. Shadows look strange and familiar at the same time; you've seen this before. It's like the time you were caught out late, in the woods, at sunset, on a cold winter evening.

Shadows are obviously sharp and well defined. You take off your straw hat, hold it a few feet off the ground and see the projected image of hundreds of tiny Sun crescents. Cool.

Act 2, Scene 2. The Company sings "Here Comes goes the Sun": The Moon is moving even faster now. Fully 95% of the Sun is covered and something strange is happening. We're in the middle of a scene out of a sci-fi movie. In the far distance, we see either a thunderstorm or a huge Martian spaceship approaching us. No—it's the shadow of the

Moon sweeping towards us at over 1,000 miles per hour. On the horizon, you notice a sunset but it's all around—a 360-degree sunset!

Thin wavy lines of alternating light and dark *shadow bands* are dancing on the ground.

With the darkness comes a sharp temperature drop of 18 degrees Fahrenheit! Glad I brought this sweater. People are excitedly animated at this point, and a few oohs and aahs are called out as the shadow engulfs the audience.

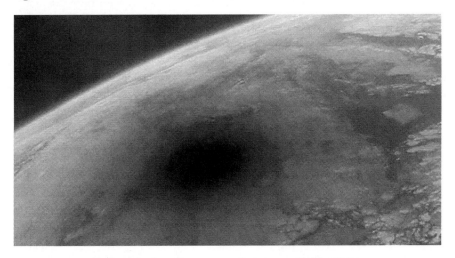

Eclipse shadow from space. Courtesy of MIR, NASA.

Act 2, Scene 3. The Company sings "Bill Baily's Beads": The Sun is determined to go down in a blaze of glory. All of its light seems concentrated into an intensely bright, incredibly thin silver sliver. Then it happens—someone begins counting out loud; "10, 9, 8, 7, 6, 5, 4 . . ." The crowd is wild with anticipation as they count the last few seconds out loud, in unison ". . . 3, 2, 1, BAILY BEADS"! At the exact instant the crowd cries "beads," the sliver breaks into dozens of incredibly intense, individual points of bright white light. These "light-beads" stand tall along the Moon's edge and appear to dance along its rim.

Act 2, Scene 4. The Company sings "Lucy in the Sky with Diamond (Ring)": Almost immediately, the twinkling pearl necklace of beads begins to shrink simultaneously from two sides and the bright light coalesces into a single large bead. A flash of light and then a bright white ring forms completely around the black disk of the Moon. The large bead and white ring paint a picture together. It looks as though a giant diamond ring has been tossed up into the sky. Someone in the crowd yells out "Diamond Ring!" A few seconds pass while the ring animates itself up in the sky like a logo at the beginning of a feature movie. Then it disappears. The sun is now totally covered by a black disk and we have reached *second contact*.

Diamond ring- Australia, November 13, 2012, by Author.

Act 2 Scene 5. The Company sings "A Corona of My Owna": The curtain opens and someone, at the top of their lungs, cries out "totality!" People around you drop whatever they were using to watch the Eclipse up to this point, aggressively ripping the protective lenses off their faces, binoculars and cameras (or telescopes).

Most people are just looking upward with their naked, unaided eyes. A few people grab the binoculars from around their necks and for the first time without special safety lenses, stare through them up into the sky. Because you read this book, you also brought binoculars so you too can enjoy a close-up look at all the skyward action.

Where the Sun used to be, there is now a jet black "hole" in the sky surrounded by a huge white halo. It is as if the petals of some incredibly beautiful cosmic flower are shooting off far into space. Many petals reach out to more than double the diameter of the Sun. We're seeing the Sun's atmosphere, the solar corona. Outstanding! The appearance of the corona is different for each eclipse.

You notice the planets Venus and Mercury in the sky, high above the horizon. Then you spot the star formations and realize the constellations are the ones you would normally see at night during the winter. They've been up there all summer during the daytime but are only making this rare appearance because of totality.

> During the totality phase of the August 21, 2017 summer eclipse, the constellation Canis Major—the greater dog—should be visible in close proximity to the Sun. The brightest star within it (and the brightest in our night sky) is Sirius, which is also known as the "dog star". Its first appearance in the dawn sky in August is why this time of year often referred to as "the dog days of summer."

Birds have stopped chirping, bees are no longer buzzing about, and an owl is making night sounds. If you're near a lake or ocean, fish may leap out of the water in search of dinner. Animals are very confused. It is nighttime. Or is it?

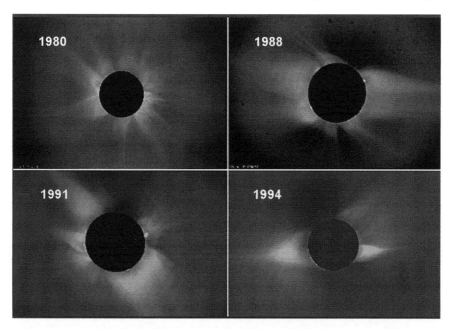

Totality reveals the constantly changing corona.

Act 2 Scene 6. The Company sings "Prominences, Prominences": Look closely along the edge of the black disc where the Sun used to be. There are small yellow-red points of light reaching tentatively up from the Sun's surface, only to fall back on themselves in small tight loops. These loops of hot gases look like miniature versions of Canyon Lands National Park rock bridges except they glow colorfully against the black of space. Just as you start to figure out how to get the sharpest image from your binoculars, you notice someone is counting out loud again. The prominences disappear from the East side of the Moon and start to show on the West side. You get the feeling that time is running out.

Act 2 Scene 7. The Company sings "Here Comes the Sun": Someone yells "SAFETY GLASSES ON." We have reached *third contact* as the Moon starts to reveal a sliver of the hidden Sun. Everyone scrambles with their eye gear, filters are slapped back on binoculars, cameras and telescopes. You make sure your kids have their eye protection back in place because totality is about to end. A bright flash of light and another diamond ring springs from the sky as the moving disk of the Moon reveals a tiny slice of the Sun. More Bialy Beads appear. Act 2 is over. The curtain closes.

The show still has one more Act to go, but the audience is too excited to care. Instead of politely waiting for the performance to finish, all rise from their seats, delivering an appreciative standing ovation.

You have just witnessed the totality phase of a total solar eclipse. A partial solar eclipse completely skips over our "Act 2" because the Moon's appearance is not large enough to fully cover the Sun.

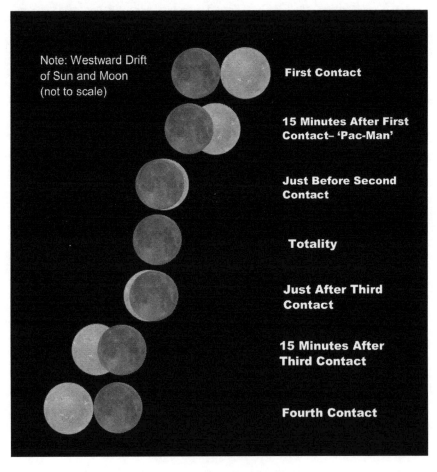

Progression of total solar eclipse.

Act 3. "The Moon Exits" (one long scene)

The Moon slowly pulls away from the sun. Act 3 will take the same amount of time as Act 1.

Act 3, Scene 1- The Company sings, "It's All Over Now, Baby Blue": People are chatting excitedly about the incredibly beautiful event they've just witnessed. Experienced observers are comparing it to other total solar eclipses they've seen. First timers are recovering from the awesomeness for which all the reading and online videos could never have prepared them. Some people are wiping away tears from their eyes. I won't admit that I cried at my first eclipse, but let's just say my wife has stories she still tells.

People begin to pack their gear even though the Moon is still covering three-fourths of the Sun. The exciting part is over. Soon the Moon will withdraw its embrace of the Sun completely. As the edges of the two discs separate, we have reached *fourth contact*. The last kiss ends and the Moon and Sun each go their separate ways until next time. The curtain closes.

Fantastic! Like Forest Gump's famous box of chocolates, every eclipse is different. Sometimes the corona is well defined, sometimes diffuse. Sometimes there are many large solar prominences, at other times only a few small ones.

Hopefully, you had the good fortune to have shared this observation experience with others. Perhaps you made a new friend or two along the way.

Now is a good time for a celebratory drink (or two). I bet you'll never look up at the sky in quite the same way again. You did it. You've accomplished what only a small number of humans have ever done. You stood in the shadow of the Moon.

Traditions: A common tradition among hot air balloonists is to enjoy a champagne toast upon landing. Legend has it that early French aeronauts carried champagne to appease angry or frightened spectators at the landing site.

For eclipse chasers, two relatively modern traditions often commemorate the successful eclipse viewing experience. In 1972 onboard the first commercial eclipse cruise, a commemorating flag was raised. That exact flag design is no longer in circulation but at the next total solar eclipse, Craig Small raised a replacement flag that has since graced thirty total eclipses and many annular eclipses in succession.

Some eclipse chasers enjoy a celebratory toast using a drink known as an "egg cream," made from a mixture of sparkling water (seltzer), milk and chocolate syrup (traditionally Fox's U-bet syrup from Brooklyn, New York). For those of you who prefer some alcoholic content, the one and only drink that contains the word "eclipse" is Mount Gay's Eclipse Barbados Rum.

Craig Small with U-bet chocolate syrup and eclipse flag on the Island of Tahiti, July 11, 2010. Photo celebrates world's longest observation of totality by the general public; totality was extended to 9 minutes and 23 seconds onboard an A319 jetliner. Courtesy of Robyn and Craig Small.

During a solar eclipse, it will appear that the Moon is making its way eastward towards its contact with the Sun. You will see the eastern edge of the Moon make first contact with the western edge of the Sun, and then the Moon will take bigger and bigger "bites" as it eats its way across.

The problem, however, is that both the Sun and the Moon rise in the east and set in the west. That is, they are both heading towards a sunset or moonset in the west. How can the Moon appear to be moving to the east during the eclipse? What we see during the eclipse doesn't make sense at all. I hate when that happens.

The explanation is simple: from our point of view, both objects are in fact moving towards their destiny to set in the west. The Moon rose before the Sun on Eclipse day but compared to the Moon, the Sun is moving faster through the sky on its way westward. The Sun catches up around first contact. The Sun is literally "passing behind" the Moon on its way to the western horizon. The western edge of the Sun catches up with the eastern edge of the Moon first. Our brain erroneously assumes the object closest to us is moving, while the object behind remains stationary.

In conclusion, while you are watching the eclipse, try to remember that the Sun is moving behind the Moon, rather than the Moon crossing over the Sun.

Educators (myself included) are often guilty of confusing this important scientific fact by our choice of language used to describe the eclipse event. You'll notice our liberal use of the word "appears" to distinguish between what you see and what celestial mechanics may have been cooking up behind the scenes.

The Sun is sliding Westward behind the Moon. A fun fact to tell people during the eclipse.

Shadow Bands: These wavy light and dark "jail-bar" like lines can sometimes be seen moving across surfaces about two-minutes before and two-minutes after totality. They are caused by the thin crescent Sun illuminating the Earth's atmosphere, with the movement driven by atmospheric winds.

Total Solar Eclipse 2017

Chapter 3. The Sun's Secrets:
The Spectacle Revealed by Totality

Baily's beads, the diamond ring, solar prominences and the corona: what are these things and how do they relate to the operation of our nearest star, the Sun? We will explore each of them in turn and along the way, you'll discover the basic operating principals of the Sun. By pulling back the curtain, you'll more fully appreciate what the cosmos is revealing on eclipse day.

Overview

The Sun is so large that it contains 99.8% of all the mass in our entire solar system! If the Sun were as tall as a typical front door, the Earth would be the size of a U.S. nickel.[9] At the same scale, the distance from the Earth to the Sun would be two football fields. The Sun is about 93 million miles from Earth and its light takes about eight minutes to reach us.

The center of the Earth consists of a part-solid, part-liquid *core*. Surrounding this is a hot, active *mantle* that includes pools of lava (melted rocks and metals). Next comes a *crust* of rock and dirt, with some valleys filled with water (oceans). Reaching 10 miles above all of this, is the portion of our atmosphere that sustains life, the *troposphere*, primarily made of nitrogen and oxygen. The point of discussing the makeup of the Earth is to emphasize how different the Sun is from our home planet.

Our Sun is a typical star; a giant ball of gas in the form of *plasma* with its core powered by a non-stop nuclear fusion reaction. While astronomers refer to the "surface" of the Sun, there is no solid surface, just a layer of gas that is opaque, appearing to us as a solid. It's an optical trick, you just can't see through to the layers of gas underneath.

"The material in the sun and its atmosphere are all *plasmas. Plasma* is a state of matter much like solid, liquid and gas. Plasmas are so incredibly hot that the electrons leave their atoms, making it essentially a gas of charged particles. While uncommon on Earth, 99% of the matter we can see in the universe is made of plasma. The electrical charge strongly affects how the particles move since the particles are simultaneously governed by, and constantly creating, magnetic fields. For example, in close-up images of solar activity you can see the plasma very clearly following the magnetic field lines. Conversely, as plasma moves, it drags its own magnetic fields along for the ride."
Source: NASA[10]

The Sun rotates about its axis; however, unlike Earth, it consists of all gases. Therefore, the Sun exhibits the unusual property of rotating about once every 25 days at the equator and slower at the poles (once every 36 days).

The Sun consists of 70% hydrogen, 28% helium and 2% other elements. It formed about 4.5 billion years ago and today is about halfway through its useful life. Every second, the nuclear fusion reaction in the core converts 700 million tons of hydrogen to 695 million tons of helium, producing 386 billion billion megawatts of energy. The energy released is in the form of radiation at various frequencies, and much of it has become visible light by the time it leaves the Sun's "surface." The heat from this reaction warms the Earth, grows our plants and ultimately makes all life on Earth possible.

The mass of the Sun is so great that it will not run out of hydrogen fuel for another 5-7 billion years. As the Sun continues its lifecycle, it will eventually expand to destroy our solar system, including Earth.

The future of Earth: In about 1.1 billion years, the solar luminosity will be 10% higher than at present. This will cause the Earth's atmosphere to become a "moist greenhouse," resulting in evaporation of the oceans. Four billion years from now, the increase in the Earth's surface temperature will cause a runaway greenhouse effect. By that point, most if not all the life on the surface will be extinct. The most probable fate of the planet is absorption by the Sun in about 7.5 billion years, after the star has entered the red giant phase and expanded to cross the planet's current orbit.[11]

The "Layers" of the Sun

The Sun's *core* exists at extreme levels of temperature and pressure: 27 million degrees Fahrenheit and a density of 150 times that of water. Put another way, a piece of the core that measured 1 x 1 x 1 foot would weigh 9,364 pounds. The same size block made of pure gold weighs 1,200 pounds.

Nuclear reactions in the core generate energy in the form of highly energized electromagnetic radiation (photons of light) that bounce around within the adjacent *radiative* layer. Once a photon of light is born, it travels a small distance until it collides with a charged particle and is diverted in another direction. This random walk is a bit like "two steps forward and one step back." Because of the extreme density of the Sun, scientists estimate the light takes at least 10,000 years and perhaps as long as 170,000 years to get from the core to the outer edge of the *radiation layer*. The energy then makes its way across a thin transition zone called the *tachocline* to reach the *conductive layer*, where the intense heat causes the gas to boil. The term "tacho," (as in tachometer) refers to the fact that the *radiation layer* and *conductive layer* are rotating at different speeds.

Next comes the 250-mile thick *photosphere*—the deepest layer of the Sun that we can directly observe. The *photosphere* appears to us as a surface pocked with fine granulation, indicating areas of varying temperature. The *photosphere* is not a solid surface, but rather a mass of hot gases that we can't see through. Temperatures at the beginning of this layer start at 11,000 degrees F and decrease to 6,700 degrees F as you move outward. All the solar effects we observe on the Sun's "surface," such as sunspots, prominences and flares appear to us as happening on the *photosphere*.

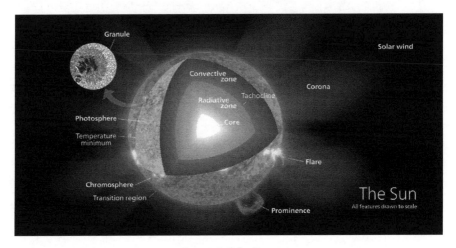

Layers of the Sun.
Courtesy of www.commons.wikimedia.org/wiki, User: Kelvinsong Sun.

The 1,000-mile thick *chromosphere* surrounds the photosphere. Temperatures start at 6,700 degrees but surprisingly, increase to 14,000 degrees F as we move outward. This layer is capped by a very thin, 60-mile skin called the transition region. During a total solar eclipse, the Moon covers the photosphere, often making the *chromosphere* visible as a red ring between second and third contact.

Finally, we reach the *corona*, the outermost layer of the Sun. Think of this less as a closely-held layer, and more like a very thin, low-density atmosphere reaching far into the expanse of space. The shape and size of the *corona* varies and it can be many times the diameter of the Sun. Astonishingly, at over 1,800,000 degrees F, it is much hotter than the Sun's lower layers. What causes the *corona* to be so hot is the subject of debate. When you look up at the sky during totality, remember that those friendly white wisps surrounding the eclipsed Sun may look touchable, but are actually over 1 million degrees!

The effects of the Sun reach far beyond the corona, in the form of a massive magnetic field called the *heliosphere* or *magnetosphere*. The Sun also emits a field of low energy charged particles (the solar wind) and high energy particles (during solar flares). These particles are mostly invisible to the naked eye, although they affect Earth by disrupting radio signals and creating the aurora borealis.

The Auroras: Energy particles from the Sun reach Earth and excite oxygen and nitrogen in our atmosphere, causing them to glow, not unlike electricity causing a fluorescent light bulb to glow. When seen near the north magnetic pole, they are called the *Aurora Borealis* or the *Northern Lights*—near the southern pole, they're called the *Aurora Australis* or *Southern Lights*. Aurora displays appear in many colors although pale green and pink are the most common.[12]

Aurora Australis from the International Space Station.
Courtesy NASA, by astronaut Joe Acaba, flight engineer (July 15, 2012).

Baily's Beads and the Diamond Ring

The Moon's surface contains mountains and valleys giving the Moon's disk (the *lunar limb*) an uneven, jagged edge. During a total solar eclipse, just before the lunar disk covers the last bit of Sun, light streaming through the valleys appears as a string of irregular size and haphazardly spaced beads. As the Moon continues to completely cover the Sun, eventually only one bead will remain. A bright white ring around the Moon appears, and the combination is reminiscent of a giant diamond wedding ring set against a black sky.

Baily's beads are named after Francis Baily who was the first to explain them in a paper to the Royal Astronomical Society, December 9, 1836:

"When the cusps of the Sun were 40 degrees asunder, a row of lucid points, like a string of beads, irregular in size, and distance from each other, suddenly formed round that part of the circumference of the Moon; and attributing this serrated appearance of the Moon's limb to the lunar mountains..."[13]

Diamond Ring. Australia, November 13, 2012, by Author.

Sun Spots, Solar Prominences, Solar Flares and Coronal Mass Ejections

During totality, red or pink features appear to cling along select spots on the lunar limb. What exactly are these *solar prominences* and how are they different from *sunspots*, *flares* or *CMEs* (*Coronal Mass Ejections*)? All four phenomena take place on, or are launched from, the "surface" of the Sun (the *photosphere*). Even the coronal mass ejections start on the photosphere, in spite of the word "corona" in the CME moniker). All these phenomena are linked, and all are driven by the Sun's intense magnetic field.

Sunspot

A *sunspot* is an area of the Sun's surface that is created by the twisting and turning of magnetic fields. *Sunspots* appear dark because they are lower in temperature than the surrounding plasma. If you could remove a *sunspot* from the intense brightness of the Sun, it would still appear as an extremely bright red glowing object on its own. They typically come and go over a period of a few days or months, and their numbers rise and fall along with the Sun's 11-year magnetic activity cycle (*the sunspot cycle*). *Sunspots* occur in pairs with opposite magnetic polarity.

Sunspots appear as dark patches on the Sun. Courtesy of NASA.

Ultraviolet profile image of sunspot loops.
Courtesy of NASA Goddard Space Flight Center.

Solar Prominence

A *solar prominence* occurs when hot electrically charged gases
(*plasma*) in the photosphere act like metal shavings between two
magnets, forming a huge loop between two opposite polarity sunspots.
These appear during the eclipse as small jets and loops of reddish-pink
streams emanating from the Sun's surface. If the sky is clear of clouds
and free of particles, prominences can appear bright red, almost
scarlet in color. These are huge hot clouds of hydrogen plasma floating
in the solar atmosphere. Some extend 50,000 miles out from the Sun's
photosphere! *Prominences* can remain for months and are
occasionally ejected far into space during a Coronal Mass Ejection or
CME (explained later in this chapter). The size of these phenomena
can be enormous as illustrated in the following composite image.

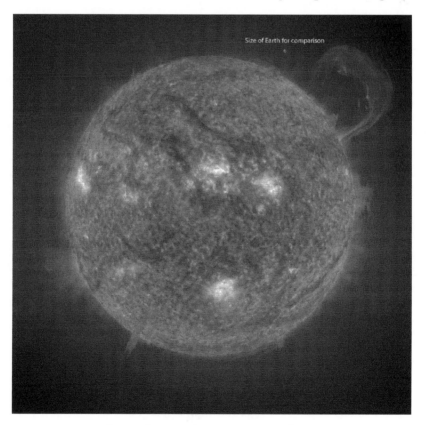

A huge solar prominence. Courtesy of NASA.

Solar Flare

During magnetic storms, a prominence can emit bursts of radiation, including visible light into space; this is a *solar flare*. A *solar flare* is often a sign that an eruption will soon take place, that is, a *CME* is about to occur.

CME (Coronal Mass Ejection)

A *CME* is usually quite spectacular. When the magnetic field lines at the base of a prominence reconnect, essentially short circuiting the magnetic field, chunks of plasma are expelled at high velocity out to space. Here on Earth, CME's can cause radio interference and are often responsible for the aurora phenomenon reaching beyond the Polar Regions into the mid latitudes.

The Corona

The *corona* is the main component of the Sun's extended gaseous outer atmosphere. In this section we will identify what it is made of, discuss its physical properties, and describe its appearance. We will also investigate the missing pieces in our understanding of the Sun.

What Is the Corona Made Of?

First we need to understand a scientific instrument called the *spectrograph*. White light contains all the colors of our visible spectrum, including red, violet and everything in between. We can separate light into its constituent colors using a prism or a diffraction grating (a plate or sheet of metal ruled with very close parallel lines).

When white light passes through a gas, atomic elements in the gas (such as hydrogen, helium and carbon) will absorb or reinforce specific colors. The *spectrograph* separates these colors and produces a chart with *spectral absorption lines* and *spectral emission lines, i.e.,* dark or bright lines for colors which are specific to the elements found in the gas.

Scientists use this technique (called *spectroscopy*) to discover the atomic make-up of almost any material, even solids. You may have seen this used in a forensic lab on a television police drama like CSI (Crime Scene Investigation). A small sample of evidence is placed in a chamber of a spectrograph. The unit heats up this sample until it emits a gas. A white light shines through the gas, and certain wavelengths (colors) are absorbed or emitted. The resulting chart reveals the elements that were contained in the sample.

As light is emitted from the Sun, it passes through the Sun's gaseous layers on its way to Earth. Early spectrographic observations of the visible spectrum of the corona revealed bright emission lines at wavelengths that did not correspond to any known materials. Prior to about 1945, astronomers explained this by saying that the solar corona was made of an entirely new undiscovered element named "coronium." This was wrong.

It was later discovered that coronal gases are superheated to temperatures over a 1.8 million degrees F (about 1 million degrees C). At these high temperatures both hydrogen and helium (the two dominant elements in the corona) and most of the remaining trace elements are completely stripped of their electrons. Gas in this state is called *ionized plasma* or *plasma*. It is these highly ionized elements that produce the spectral emission lines that were so mysterious to early astronomers.

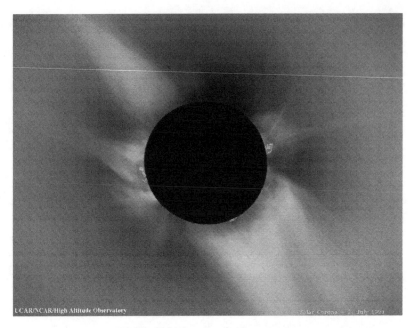

Visible light image of solar corona.
Courtesy of NASA Marshall Space Flight Center Solar Physics.

Close up of a loop of the corona arising from the photosphere.
Courtesy of NASA.

The Corona's Properties

We already mentioned the startling, stand-out feature of the solar corona: It has a temperature of over 1.8 million degrees F, 100 times hotter than the surface of the Sun! Unlike the photosphere and chromosphere, whose relative thicknesses are comparable to the skin of an orange, the corona reaches out into space up to 10 times the diameter of the Sun. The inner portions of the corona are about as bright as the Full Moon and during totality can be seen by the human eye as far out as two to three times the Sun's diameter.

Sunlight reflecting off the charged particles of gas (the plasma) makes the corona visible, however these gases are spread over a large distance and are very low in density. The Sun's constantly changing magnetic forces sculpt the plasma into the cloud-like wisps and flower petals we see during totality.

Because the further-out coronal atmosphere density is so low, X-ray detectors are used to visualize these regions.

Mysteries and Ongoing Research

Several unanswered questions about how the Sun operates remain: How are the huge magnetic fields generated and how does the corona get so hot?

The current theory is that the magnetic fields are generated by rapidly circulating plasma in the tachocline layer. In this model, energy is transferred from the magnetic field to the coronal plasma. The following is an excerpt from a NASA website:

"The magnetic field is dragged around and twisted by the turbulent motions of the gas in the convection zone. These motions can propagate up the magnetic field lines to the corona, where the magnetic field pressure is greater than the gas pressure. There the extra energy transferred to the magnetic field is presumably transferred to the coronal plasma. We do not know how the energy in the magnetic field is converted to heat in the corona. One possibility is that the energy is released in many tiny flares, each too small to be observable. Another possibility is that the energy is in the form of hydromagnetic waves that are damped in the corona. The study of mechanisms for transporting and depositing energy into the corona is an active area of solar research."[14]

Solar and Heliospheric Observatory (SOHO)

SOHO is a joint effort spacecraft project between NASA and the ESA (Europe Space Agency) to study the Sun, from its deep core to the outer corona and the solar wind. SOHO was designed to answer the following questions:[15]

What are the structures and dynamics of the solar interior? Why does the solar corona exist and how is it heated to its extremely high temperature? Where is the solar wind produced and how is it accelerated?

SOHO was built by European contractors and launched by NASA in 1995. It is operated by NASA's Goddard Space Flight Center (GSFC) near Washington with ground control provided by NASA's Deep Space Network antennae, located at Goldstone (California), Canberra (Australia) and Madrid (Spain).

The SOHO spacecraft remains in orbit today. It contains 12 observational instruments including the LASCO (Large Angle and Spectrometric Coronagraph). LASCO produces an artificial eclipse by artificially blocking out the center image of the Sun and photographing the surrounding corona.

SOHO revealed the first images of a star's convection zone (its turbulent outer shell) and imaged structural details of sunspots below the surface. It gave us the first precise measurements of temperature, interior rotation rates, and gas flows in the Sun's interior and within the solar wind. SOHO also revealed several new dynamic solar phenomena such as coronal waves and solar tornadoes. Below are images from SOHO.

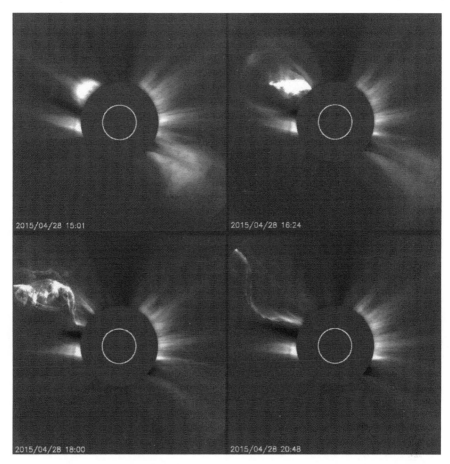

Image sequence from LASCO onboard the SOHO spacecraft. An elongated solar filament that extended almost half the Sun's visible hemisphere erupted into space (Apr. 28-29, 2015) in a large burst of bright plasma. The LASCO is a "coronagraph" which produces an artificial eclipse by blocking out the center image of the Sun and photographing the surrounding corona. The white circle in the center of the round disk represents the size of the Sun, as delineated by its Photosphere, which is being blocked by the occulting disks. Everything past this white ring is happening in the Sun's huge corona. Credit: SOHO, NASA/ESA.
http://sohowww.nascom.nasa.gov/pickoftheweek

Solar Dynamics Laboratory (SDO)

NASA launched the Solar Dynamics Laboratory in 2010 and today it remains in geosynchronous orbit around the Earth over New Mexico. Its mission is to help us understand the Sun—Earth system and how it affects life and our society. SDO studies the Sun's magnetic field, the solar wind and space weather here on Earth.

Solar Probe Plus

"A NASA Mission to Touch the Sun."

The next major NASA solar research craft, the Solar Probe Plus, is under development with the official program schedule calling for a launch in July 2018. In NASA's own words:

"Solar Probe Plus will be a historic mission, flying into the Sun's atmosphere (or corona) for the first time. Coming closer to the Sun than any previous spacecraft, Solar Probe Plus will employ a combination of in situ measurements and imaging to achieve the mission's primary scientific goal: to understand how the Sun's corona is heated and how the solar wind is accelerated. Solar Probe Plus will revolutionize our knowledge of the origin and evolution of the solar wind."[16]

The Johns Hopkins University Applied Physics Laboratory (APL) is developing the mission, which will study the streams of charged particles the Sun hurls into space from an unprecedented vantage point: inside the Sun's corona. The 1,350-pound craft (about half a car in weight) is protected by an 8-foot diameter, 4.5-inch-thick carbon-composite heat shield.

Solar Probe Plus will orbit the Sun 24 times, gradually walking inward with each pass. The spacecraft will fly to within 8.5 solar radii of the Sun's "surface" (about seven times closer than we have been before) and well within the orbit of Mercury, the closest planet to the Sun. Perhaps we will finally solve the mystery of what drives the corona to such an incredibly high temperature.

Chapter 4. Up There:
How Eclipses Happen

The dictionary defines an astronomical eclipse as "the total or partial obscuring of one celestial body by another."[17] This broad definition leads us to the two general types of eclipses commonly observed here on Earth: the solar eclipse and the lunar eclipse. This book is exclusively focused on solar eclipses. However, a sidebar is included later in this chapter to explain the basics behind a lunar eclipse.

> Solar eclipse: The Moon's shadow falls on part of the Earth.
>
> Lunar eclipse: The Earth's shadow falls on part or all of the Moon.

Solar Eclipse Basics

Light from the Sun reaches across space and strikes the Moon. Just like any ordinary object, the Moon casts a shadow into space. When the Earth happens to be under this shadow, we get an eclipse of the Sun (a solar eclipse).

To illustrate, we'll use a "thought experiment." As you read what follows, try to picture it in your mind's eye.

Remember, it is the Moon's shadow that causes a solar eclipse: Picture a flashlight (the Sun) shining towards your face from the far side of the room. Your head is the Earth. Now, imagine holding a baseball in your hand and extend your arm straight out towards your Sun. That Baseball is the Moon. If you get perfectly lined up, a shadow will be cast by the baseball (the Moon) onto your head (the Earth). That is an eclipse of the Sun.

Notice that during a solar eclipse the Moon and the Sun are more-or-less lined up on the same side of the Earth. We call this a *New Moon*. A solar eclipse can only occur when there is a *New Moon*. However, we don't have a solar eclipse with every *New Moon*. This is because, for most *New Moons*, the Moon's shadow is cast at an upward or downward angle, missing the Earth.

Bonus astronomy term "syzygy": A cool word, not used in the text to keep things simple. The term refers to the alignment (in three dimensions) of three celestial bodies such as the Sun, Moon and Earth in, or nearly in, a straight line such as during an eclipse. Pronounced "scissor-gee." "The Sun, Moon and Earth will be in syzygy during the eclipse."[18]

How a lunar eclipse works: When the light from the Sun strikes the Earth, it casts a shadow. If the Moon falls within this shadow, we get a lunar eclipse. Picture a flashlight (the Sun) pointed at you from across the room. Your head is the Earth, and you are sitting in a swivel chair. Now, hold the baseball (the Moon) in hand and extend your arm straight out towards your Sun. Next, spin your chair slowly until you are facing the exact opposite direction; you are looking away from the flashlight. If you get perfectly lined up, there will be a shadow cast by your head (the Earth) onto the baseball (the Moon). That is an eclipse of the Moon.

Notice that the Moon is on the opposite side of the Earth from the Sun. This is a *Full Moon*. A lunar eclipse can only occur when there is a *Full Moon*. However, we don't have an eclipse with every *Full Moon*. Just like the solar eclipse situation, for most *Full Moons*, the Earth's shadow is cast at an angle, missing the lunar surface.

Looking at this another way, during a Full Moon, we on Earth are facing away from the Sun and therefore it is our night. Under normal (non-eclipse) conditions, the light from the Sun passing under or over the Earth, strikes the lunar surface. The Moon makes its scheduled monthly appearance as our fully illuminated disk in the romantic night sky. Occasionally Earth's shadow lines up with the Moon and we observe a lunar eclipse.

Why does the Moon appear red-orange during a lunar eclipse? During a lunar eclipse, the Full Moon is in the Earth's shadow, and you might expect it to appear black. However, sunlight strikes the back and sides of Earth, is refracted, bends around the planet, passes through our atmosphere and heads out into space. On its trip through our atmosphere, the blue portion of the light is absorbed and the resulting red *earthshine* reflects on to the surface of the Moon, painting it red.

Why the Moon's Shadow Usually Misses Earth

As you know, our Moon orbits the Earth approximately once a month, therefore a New Moon occurs once every month. That's where the term comes from; the Moon starts a new monthly cycle. Why then, don't we have a solar eclipse every month?

The reason is that the Moon's orbit around the Earth is tilted relative to the orbit of the Earth around the Sun. Most of the time this tilt causes the Moon's shadow to project into space, either too high or too low to hit the Earth.

Think of it like this: In an eclipse, all three objects (Sun, Moon and Earth) must be more-or-less aligned in three dimensions.

Most of us imagine the orbits of the Sun, Earth and Moon as circles (or ellipses) on a flat piece of paper. But this places them all on the same plane. That is not how it works. As Mr. Spock reminded Captain Kirk, you have to think in three dimensions.

> The dangers of two-dimensional thinking. From the 1982 movie, "Star Trek II: The Wrath of Khan": During a battle, Kirk directs the Enterprise into a nebula where static discharges render shields useless, making the Enterprise and the ship Khan is commanding evenly matched. Spock advises Kirk that Khan's moves "indicate two-dimensional thinking," leading Kirk to exploit Khan's inexperience by ordering Enterprise to drop 10,000 meters and then attacking Khan's ship from below.[19]

If you build a Styrofoam model and animate the orbital motions in three dimensions, you will find that, because of tilt, the orbit of the Moon around the Earth crosses the plane containing the orbit of the Earth around the Sun twice each year. When this happens AND there is a New Moon, we get a solar eclipse. Chapter 10, "The Motions of the Sun, Earth and Moon" explains eclipse mechanics in more detail.

> Bonus astronomy term, *"ecliptic" (noun):* The plane of Earth's orbit around the Sun is called the *ecliptic*, not to be confused with the word *elliptic* or *elliptical* (the shape of the orbit). The Moon's orbit around the Earth is tilted about 5 degrees relative to the *ecliptic*.
>
> Origin of the word *ecliptic*: "the circle in the sky followed by the Sun," from Medieval Latin ecliptica, from Late Latin (linea) ecliptica, from Greek ekliptikos "of an eclipse." So called because eclipses happen only when the Moon is near the line.[20]

Types of Solar Eclipses: Partial, Annular and Total

In ranking the excitement of observing each type of solar eclipse (from 1 to 10, with 10 being the best) I rank the *partial solar eclipse* a 1 (pretty dull), the *annular solar eclipse* a 4 (interesting but a bit of a yawn) and the *total solar eclipse* a clear 10. What is the difference between these eclipses and how is each one created?

It may go without saying, but for clarification, don't confuse the term *annular* (ring-like) with annual (once per year). An *annular* eclipse does not happen once per year! The term refers to the fact that a ring (an annulus) of the Sun can still be seen during the eclipse. An *annular eclipse* appears much like a *partial eclipse* with the sky never becoming dark, and the Sun's corona remaining blocked out by the brightness of the remaining solar ring.

Why Do Solar Eclipses Happen at All? Think 400.

The Earth-Moon-Sun geometry is different for each eclipse type, but the same celestial coincidence makes them all possible. The Sun is about 400 times the diameter of the Moon. The distance from the Earth to the Sun is about 390 times further than the distance from the Earth to the Moon. Due to this happy accident and the physics of optical perspective, the relatively tiny Moon can appear the same size as the huge Sun. From the perspective of a viewer on Earth, this allows the Moon to block out the solar disk.

By the way, our Moon measures only about 27% of Earth's diameter, around 2,159 miles across. That's the same distance as from Boston to Denver. The Sun is huge, about 109 times the diameter of the Earth.

The Moon's Shadow Determines the Type of Solar Eclipse

The shadow cast by the Moon actually contains three different shadow portions clustered together. The *umbra* is the "cone-shaped" dark center shadow; the *antumbra* is the shadow that takes over where the *umbra* ends; and the *penumbra* is the lighter shadow that surrounds the *umbra* and the *antumbra*. The type of eclipse that occurs is determined by which part of the Moon's shadow reaches Earth. Take a moment to study the illustration that follows.

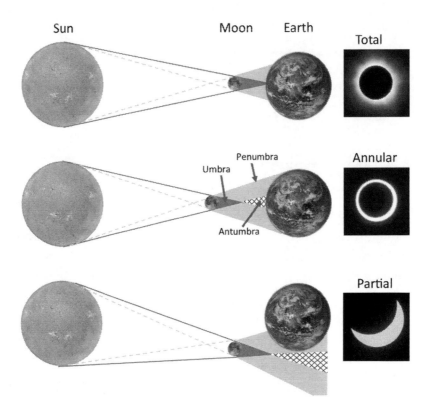

Three types of solar eclipses: The umbra is dark gray, penumbra is light gray and antumbra crosshatched in red. Not to scale.

Three different types of solar eclipses are possible:

1. *Total Solar Eclipse:* The *umbra* and *penumbra* reach Earth (see the above figure). When you stand under the *umbra*, the Moon appears to cover the Sun completely. The Sun appears only partially blocked when you are under the *penumbra*.

2. *Annular Solar Eclipse*: The *antumbra* and penumbra reach Earth (see the above figure). However, the Moon's apparent size is not large enough to cover the Sun. When you stand under the *antumbra*, the eclipse appears as a bright white ring with a dark disk (the Moon) in the center. The Sun appears only partially blocked when you are under the penumbra.

3. *Partial Solar Eclipse*: Only the penumbra reaches earth (see the above figure). The Sun appears only partially blocked when you are under the penumbra.

The inner portion of the shadow called the *umbra* is dark and focused. The *umbra* is the diameter of the Moon when it leaves the lunar landscape, and like an ice cream cone, shrinks rapidly as it extends into space. At the bottom of the cone, it comes to a pencil tip and then ends. From there, a reflection of the *umbra* begins; it starts out as a point and becomes larger as it reaches further into space (an upside down ice cream cone). This is the *antumbra*, meaning the anti- or opposite of the *umbra*. Think of it as a mirror reflection of the *umbra*.

The third, outer portion of the shadow is called the *penumbra*. The *penumbra* is not very dark and appearing diffuse as if it's being washed out by a bright lamp from a distance. The *penumbra* starts out the same diameter as the Moon (2,159 miles), but instead of coming to a point, it gently spreads out as it reaches further into space. By the time the *penumbra* reaches the Earth, it can be as large as 3,728 miles in diameter, about half the size of Earth.

The terms umbra and penumbra are not exclusive to eclipses, but terms used in the physics of light. The umbra is the portion of a shadow where all of the light from the source is blocked. The *penumbra* is the portion of the shadow where not all of the light has been blocked. For a *point-source* (a laser for example) only the umbra is cast. When a larger, *non-point-source* illuminates a smaller object, both types of shadows are cast. The full Sun is not a *point-source*.

Here is how it works: Imagine a shade lamp (not a point-source) casts the shadow of an apple on a wall about two feet away. The rays from the center of the lamp shade that are blocked by the apple will cast a shadow in one place on the wall. The rays from the edge of the lamp shade will cast a shadow in a slightly different place on the wall. It is a little like having two separate focused point-sources of light, but they are separated by a distance. The shadows are going to land at two different locations on the wall. In reality, we have millions of light-rays all coming from slightly different places off the lamp shade, striking the apple at varying angles. Light-ray A (from the center) is blocked by the apple and does not contribute to lighting the wall in place C. But light-ray B is coming from the edge of the shade, and it passes around the apple, adding light to location C. The result is a fuzzy shadow. In the center of the apple's shadow, there is an area that is not being illuminated by these off-axis rays. This is the umbra.

To summarize, the following conditions give rise to a solar eclipse:

IF... The Sun and Moon are lined up on the same side of the Earth (a New Moon),

AND... The Moon, in its tilted orbit, happens to be crossing the ecliptic plane (the plane of the Earth's orbit around the Sun),

THEN... One of these will occur: a total solar eclipse, an annular solar eclipse or a partial solar eclipse.

> *Hybrid solar eclipses.* There is a fourth type of very rare solar eclipse termed a "*hybrid*" or "*annular-total*" *eclipse.* A *hybrid* is the same as a *total solar eclipse,* except that at some place on the Earth's surface, it appears as only an *annular eclipse.* This happens because the Earth is curved, and as the Moon's shadow sweeps across the globe, for some area under the *umbra,* the ground has "fallen away." In these geographies, the Moon's shadow does not reach all the way to the surface with the result that the observer sees only an *annular eclipse.* The duration of totality during a hybrid eclipse is usually very short.

It is important to note that during a total eclipse, both the umbral and penumbral shadows reach the Earth. However, you will only see totality when you are under the umbra. The umbra is where eclipse chasers and people in the know (like you) travel to when a total eclipse is coming.

The general public often misses totality because they don't realize that they must be under the narrow path of the umbral shadow.

The situation is similar for an annular eclipse: You must be under the antumbral shadow to see it as a ring. Outside this, people see it as a simple partial eclipse.

What Determines the Type of Solar Eclipse?

To understand how each different type of eclipse comes to be, we need to consider one final aspect of Moon-Earth-Sun geometry.

The Moon's shadow is always present, it's just a question of whether it hits the Earth. The penumbral shadow spreads wide by the time it reaches the distance to Earth. Therefore, even if the Sun, Earth and Moon are only approximately aligned, there is a good chance the penumbra will graze our planet and we get a partial eclipse.

On the other hand, an annular and a total eclipse depend on a shadow that becomes very small by the time it reaches the general area of the Earth (sometimes as small as 50 miles in diameter). This means the alignment must be precisely right. The Sun, Earth and Moon must arrange themselves with their centers lined up precisely. For this reason, we refer to annular and total eclipses as *central eclipses.*

What determines whether a *central eclipse* is an annular or a total? Recall that during a total solar eclipse, the apparent size of the Moon is about the same as the Sun. However, since the orbit of the Moon is not a circle but is elliptical, the Moon will appear smaller when it is further away in its orbit. When it appears too small to totally cover the Sun, we get an annular eclipse. When the Moon is closer to us, it can cover the Sun, and we get a total eclipse. The Moon's apparent diameter varies by about 12% throughout its orbit.

The Earth's elliptical orbit (once per year) around the Sun also causes the apparent relative size of the Sun and Moon to change; by about 3%. The effect of Earth's orbit around the Sun is much smaller than the 12% contribution from the Moon's eliptical orbit.

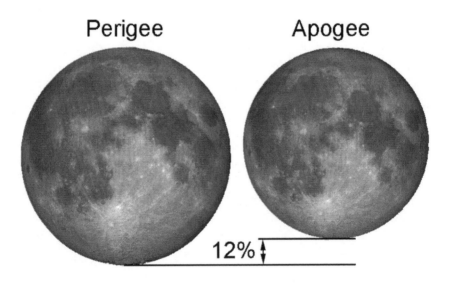

Perigee   Apogee

12%

The portion of the Moon's elliptical orbit that is furthest away from Earth is called apogee. In contrast, perigee occurs when the Moon is closest to the Earth. At apogee, the Moon is not large enough to completely cover the Sun. The result can be an annular eclipse instead of a total eclipse.

Lunar Orbit • Earth Orbit • Moon's Shadow • Umbra • Moon • Sunlight • Earth

Almost all eclipse diagrams show the Moon much closer to Earth than reality. This correct-to-scale diagram gives a better feel for how fortunate we are to have a total solar eclipse on Earth. Note the tiny umbral part of the Moon's shadow as it is about to sweep over the globe.

The dark side of the Moon: The elliptical shape of the Moon's orbit also affects how much of the lunar surface we see here on Earth.

We don't see the back side of the Moon because gravitational effects synchronized the orbit of the Moon around the Earth and the rotation period of the Moon on its own axis. Long ago, the Earth's gravitational force deformed the Moon into a shape that is not perfectly round. The Earth's gravitational field pulled a tiny bit harder on the portion of the Moon that has higher mass, and this caused the Moon to spin on its axis. When equilibrium was reached, this natural synchronization resulted in the same face of the Moon always pointing towards Earth.

If that were all that was happening, we would always see the same half (50%) of the Moon's surface, but it's not that simple. To observers on Earth, the Moon appears to rock back and forth over time. This lunar *libration* has three principal causes: 1) The Moon's orbit is elliptical so there are times, (for example, when the Moon is further away), when there is a slight change of its orbital speed. This speed change, while the Moon's rotation on its axis remains constant, causes the portion of the lunar surface that faces us to shift back and forth a small amount. 2) The Moon's rotational axis is tilted slightly relative to its orbit around the Earth. 3) Since we are standing on the Earth's surface and not at its center, each day we shift our view from left to right of the Moon's center, allowing us to see a small amount around the sides. Because of these phenomena, over time the Moon exposes 59% of its surface to observers on Earth.

Just to be technically precise, there is no such thing as the "dark side of the moon." All sides of the Moon at some point will receive light from the Sun. A more appropriate term for this hidden lunar real estate is the "far side of the Moon."

During NASA lunar missions of the 1960s, the term "dark side of the Moon" referred to the portion of the lunar orbit that sits in radio silence.

> *Dark Side of the Moon* was also the name of Pink Floyd's eighth studio album, released in 1973. While there was no song by that name, the last track on side two was named "Eclipse."

Headline News: Total Solar Eclipses Soon to Be Extinct!

Well, "soon" is a relative word. The Moon formed about 4.5 billion years ago. Since then, it has been moving away from Earth at the rate of about 1.6 inches each year. In about six hundred million years the Moon, even at the closest point in its elliptical orbit, will appear too small to completely block out the Sun.[21] At that point, the only type of central eclipses visible on Earth will be annular eclipses. No more totals; it's a limited engagement, so you had best see it "soon."

> Tidal acceleration: This complex phenomenon is why the Moon is moving away from the Earth by about 1.6 inches each year, eventually causing total eclipses to become extinct.
>
> Flowing portions of the Earth's massive oceans and interior lava pools, slosh around, creating frictional losses and causing an uneven distribution of mass. This causes the Moon's gravity to pull unequally on the Earth, creating a slight bulge at the equator. The resulting gravitational pull by the Moon on the slightly out-of-round Earth, exerts a rotational torque (force) slowing the Earth's rotation and increasing its distance from the Moon.
>
> The slowing of Earth's rotation about its axis also increases the length of the day. On average, our day is increasing by about 2.3 milliseconds (0.0023 seconds) every 100 years.

By the way, we enjoy a first class ticket to a show that doesn't play out on any other planet in our solar system. For example, although astronauts can stand on the Moon and watch the Earth block out the Sun, the Earth appears much larger in the Moon's night sky than the

Moon appears to us here at home. After all, the Earth is about five times the diameter of the Moon. The result is that during a total solar eclipse as seen from the Moon, the huge apparent size of the Earth not only blocks the light from the Sun, but also hides the Sun's corona. There are over 150 moons in our solar system. Our Moon is the only one that appears from its parent planet, to be about the same size as the Sun.[22]

Although Earth is the only <u>planet</u> that has a perfectly fitting total solar eclipse, other eclipses occur in the solar system. We know of one other place that experiences a tight-fit total solar eclipse similar to what we experience on Earth: the moon Europa, which orbits Jupiter.

Jupiter has 62 moons, including four that are large (Europa, Callisto, Io and Ganymede). If you stand on the moon Europa, the moon Callisto will occasionally pass in front of the Sun, delivering a total solar eclipse similar to what we see on Earth. Due to the large distance from the Sun to Jupiter, the eclipse objects appear smaller and the event is not as spectacular as the one here at home.[23]

Unusual photo captured by Apollo 12 astronauts from space on the way back from the Moon. The Earth has moved in front of the Sun, creating an eclipse. On November 14, 1969, just four months after Apollo 11 carried the first human to the lunar surface (Neil Armstrong), the second lunar landing mission, Apollo 12 was launched. On board were Charles "Pete" Conrad, Alan L. Bean and Richard F. Gordon. They carried the first color television camera to the lunar surface, but the transmission was lost after Bean accidentally destroyed the camera by pointing it at the Sun. Courtesy Apollo 12, NASA.

Total Solar Eclipse 2017

Chapter 5. Down Here:
How the Shadow Moves Across the Earth

There is a total solar eclipse about once every 16 months. However, only a small percentage of the population has seen this awe-inspiring event. In fact, it is quite possible that out of all your friends, no one has ever seen one. If total eclipses are this frequent, why have so few people seen one? It is because by the time the Moon's umbral shadow reaches Earth, it is less than 100 miles in diameter. Most people live outside of the umbra, and never bother traveling to see totality.

It can be confusing to the public. Here's what happens: Your local TV station mentions there is a total solar eclipse coming and that it will be visible in your hometown of Portland. What they don't fully explain for example, is that the total eclipse is only "total" in Seattle. A "total solar eclipse" is going to occur, and a part of it is going to be visible in Portland. Unfortunately, it's the boring part. People are often not even aware they missed the big show under the umbra in Seattle.

Shape of the Shadow Path

Because the Earth is spinning on its axis (one revolution per day), the eclipse shadow will appear to carve a path across the Earth's surface. Imagine a laser light pointed at a spinning ball. Due to the direction of the Earth's rotation, the shadow will always sweep more-or-less from west to east. I say more-or-less because the path will be curved to an extent depending upon how far away from the equator the shadow lands. The resulting shadow path can take on a strong north-south component superimposed over its general movement eastward.

The August 21, 2017 path of totality starts in the North Pacific Ocean, roughly 2,400 west of the Oregon coast. It sweeps from Oregon to South Carolina in about 65 minutes. It then heads southeast about 3,600 miles past the U.S. Coast, ending in the Atlantic about midway between South America and Africa. The shadow will cover 8,640 miles in 3 hours and 13 minutes at an average speed of 2,686 miles per hour. That's 3.5 times the speed of sound!

(Note: There is a more in-depth discussion of what determines the speed and direction of the eclipse shadow in Chapter 10, "The Motions of the Sun, Earth and Moon.")

Upon how much of the Earth's surface will the umbra fall? The diameter of the umbral shadow depends on the relative apparent sizes of the Sun and Moon on the day of the eclipse. For the total solar eclipse of August 21, 2017, the Moon will only be a small amount larger than the Sun and, therefore, the umbral shadow will be fairly small: about 60 miles in diameter. The umbra can reach as large as 170 miles across for some eclipses.

In contrast, the path the shadow will sweep across is large: from the Pacific to the Atlantic Oceans. The geography covered by the shadow depends primarily on where the eclipse falls in latitude (position north-south). Do you remember that latitude and longitude lesson from school?

To recall which is which, I think of the Jimmy Buffet song, "Changes in Latitude." Jimmy sings about needing a change in *attitude* (getting more laid back) by changing his *latitude*. This reminds me that latitude measures where we are on the Earth's surface relative to the equator. Move away from the equator (higher latitudes) we get a colder, more uptight climate. Near the equator, it's tropical and laid back. So Jimmy wants to change his latitude to warmer climates, presumably by moving south from the chilly parts of the United States.

If other things are all equal, an eclipse will sweep more of the Earth's surface when the shadow touches down near the equator than if it touches down in higher latitudes. This outcome is intuitive. Imagine that the shadow first touches down near the North Pole, in Northern Greenland. It would skip back into space quickly as there is not a lot of surface for the shadow to hit before space starts again. Now imagine the shadow initially touches down somewhere along the equator. There is plenty of Earth to rotate underneath the shadow. It will cover substantially more geography before skipping off into space.

Eclipse Location

There are no geographical "magnets" for solar eclipses; they are widely distributed across the globe. Astronomers, such as Fred Espenak, have predicted every eclipse that is going to happen for the next 3000 years. See Espenak's "Five Millennium Catalog of Solar Eclipses" [-1999 to +3000], available on NASA's eclipse site at:

http://eclipse.gsfc.nasa.gov/SEcat5/SEcatalog.html

Eclipse Time of Day and Duration

Solar eclipses can occur at any time of the day from sunrise to sunset. What this means on a practical basis, is that if the shadow reaches your location early, you may need to get up, travel and setup before daybreak. The longitude (east-west position) from which you plan on observing will affect what time the eclipse starts. Tables are published on the Internet specifying first contact and other important event times for upcoming eclipses. Chapter 6, "Planning the Trip," explains how to interpret the published timetables. Chapter 12, "Resources" lists websites with eclipse timetables.

The closer the Earth is to the Moon, the larger the umbral cone diameter that touches down, and the longer totality will last. The theoretical maximum length of totality of a solar eclipse is about seven-and-a-half minutes, typically it is between two and five minutes. While the full eclipse experience will last about three hours (largely because of the spinning Earth) the exciting part, totality, passes by very quickly. This has practical implications as you want to be in place and prepared to fully appreciate every second of the main event. Totality for the August 21, 2017 eclipse will vary between about 2 minutes, 0 seconds and 2 minutes, 40 seconds for U.S. land locations along the centerline of the umbral path. Chapter 7, "Choosing a Viewing Site" discusses this in more detail.

Is the Moon up during the day or only at night? It's a common misconception that the Moon is only up in the sky at night even though most of us at one time or another have noticed the Moon during daylight hours. The Moon is "up" (above the horizon) for 12 hours of every 24 hours. These 12 hours may be during our night time or may be during our daytime. Usually this up-time is split between nighttime and daytime. On average the Moon is above the horizon in the daytime sky for six hours each day. When the Sun and Moon are both high in the sky, the Sun's brightness prevents us from seeing the Moon.

Eclipse Elevation (Altitude)

During an eclipse, the Sun can be directly overhead (high in the sky) or barely above the horizon (low in the sky). This is an important factor for observers and is referred to as *elevation* or *altitude*.

The Sun is said to be at 90 degrees elevation when it is overhead, at its highest point or *zenith* in the sky. When the Sun is just above the horizon, it might be, for example, at 10 degrees elevation. Although the height of the Sun in the sky is measured by elevation, to fully specify the location of the Sun, the angular position relative to north, or *azimuth*, is also required. For eclipse watchers, elevation is more important, as will soon be explained. We mention *azimuth* for completeness.

> *Azimuth*: The measurement of the Sun's location relative to the Earth's true north. It's the same as a compass heading. Think of a clock face painted on the Earth's surface with you at the center and the "12 o'clock" position lined up with the North Pole. Call the 12 o'clock position *zero degrees*. At the 15-minute mark is *90 degrees*. If the Sun is due east of your position, it is said to be at 90 degrees. If the Sun is due west, it is at 270 degrees and so on.

It is easy to find a good observation site for eclipses that occur high in the sky. An eclipse low on the horizon requires more careful site selection. If the eclipse is very low in the sky, you may need to find a site without obstructions, on a flat open plain. Your visibility can be easily blocked by mountains, trees, buildings, etc.

Also, mid-day, high-in-the-sky eclipses usually have better weather prospects. With an early morning, low elevation eclipse, you can run into fog. Late in the day eclipses have a higher probability of coinciding with afternoon thunderstorms.

On the plus side, a low elevation eclipse can be a once in a lifetime opportunity to create a unique photograph that might be impossible with a high Sun. Try capturing the eclipse and a beautiful natural landmark in the same frame. Also, the same way a sunset produces vivid colors, a low elevation eclipse often results in increased saturation of the red and yellow light coming from the photosphere and prominences.

When an eclipse is positioned very low on the horizon (5 degrees or less elevation) and is also over a body of water, additional effects can add to the beauty of the scene. A beam of the reflected coronal light might be visible on the surface of the water or perhaps an entire Moon-Sun reflection will appear to reach up and meet the actual eclipse.

The "Moon Illusion": This is an optical illusion which causes the Moon to appear larger when it is near the horizon than it does when it's higher up in the sky. The Moon is not actually larger when it is near the horizon. You may recall seeing an almost cartoonish incredibly huge Moon at some point in your life; that is the illusion we are talking about.

Books have been written about this phenomenon and there is still controversy about what causes the effect. One of the best explanations is the "relative size hypothesis." It states that the perceived size of an object depends upon the size of objects in its immediate visual environment. The detailed objects on the horizon (buildings, trees, etc.) make the Moon appear large, however when the Moon is high and surrounded by large expanses of empty sky, it appears smaller.

The effect is created purely by our brain, not the optics of our eyes.

Chapter 6. Planning the Trip:
What to Bring and Other Tips

Planning to observe a total solar eclipse is a bit different than putting together your run-of-the-mill road trip or flying vacation. You will need to arrive at a very narrow track of land, at a specific time of the day while possibly a hundred other people are heading for that exact same place, at the same time. The spot you are heading for will not have a road sign saying "Eclipse Parking Here."

Another difference is that clear (or mostly clear) skies on the day of the eclipse is essential. To assure this, you will be checking the weather the day before, and rechecking it early the morning of the eclipse. The last difference between this and other trips is that before you go, you will need to purchase eye protection, and possibly camera filters.

In this section, we prepare for eclipse day. We'll cover what things you need to consider while planning the trip, what special items to purchase in advance, and what other things to bring along. The chapter after this one (Chapter 7, "The Best Places to Go on Eclipse Day") will help you select your viewing site.

Not Planning (Or How to Ruin an Eclipse)

The good news is that a partial eclipse will be visible from any place in the continental U.S., Mexico and most of Canada on August 21, 2017. However, you can't just walk out your door the day of the eclipse and look skyward. It won't reward you in the slightest, and you will have missed the important part of the event: totality. Unless you happen to live under the umbra, to witness totality you will need to plan a trip.

At lunch the other day, I managed to get my 86-year-old father excited about seeing the eclipse. I could see the disappointment on his face when he realized that viewing totality would require a trip away from home. Staying local, Dad would only be able to see the partial phases of this eclipse.

Even if you decide to stay home, a lack of planning can be dangerous. Many people think they can simply step out the front door and view the eclipse—without any preparation. Safely observing the partial phase of the eclipse requires you purchase proper eye protection in advance.

What is the easiest way to ruin the experience of seeing the total solar eclipse on August 21, 2017? Answer: Not planning a trip to the umbral shadow! You will end up seeing a partial eclipse, while your friends who traveled to the umbral shadow will experience the whole enchilada.

Organized Trips on Land

Before you take on the task of pulling together this trip on your own, consider joining a professionally organized group tour instead. I did this for my first trip and loved it. You get the benefit of someone else making the big decisions about observation site, transportation and lodging. You also share the camaraderie of fellow eclipse chasers. These tours often include pre-eclipse astronomy lectures and on-site guidance by experts on the big day. If this will be your first totality event, it's nice to have someone narrate the eclipse, pointing out phenomena in real-time.

A tour operator will have access to multiple weather forecasting experts and data sources. If necessary, they will shoulder the responsibility of moving the group to an alternate site.

To find a group tour, search the Internet for "eclipse tours" and see Chapter 12, "Resources." You can also inquire at your local astronomy club. *Sky and Telescope* magazine usually conducts an organized tour, and both the online and print publication contain advertisements by tour operators.

Viewing an Eclipse at Sea

What about an eclipse cruise? Before we talk about this, I need to point out that the August 21, 2017 Eclipse is a rare opportunity to see totality on dry land, right here in the United States. For this particular trip, driving or flying to an eclipse site will be inexpensive and convenient. However, for other eclipses, traveling and observing from a cruise ship can be a simple, hassle-free, and relatively inexpensive option.

On July 8, 1972, the world's first eclipse cruise departed from New York's Pier 97, aboard Greek Line's ship, the Olympia. Marcie Sigler, Phil Sigler and Ted Pedas, launched "Eclipse '72: Voyage to Darkness" with 834 passengers and one cat on board. The organizers initially had a difficult time convincing any cruise line to sign up but Marcie's two years of persistence paid off when ailing cruise company, The Greek Line, signed up. Apparently Marcie's ability to speak the Greek language helped cement the deal.

The ship positioned itself 900 miles out in the North Atlantic, under a "hole" in the sky, avoiding the inclement weather that clouded out land based observers. Onboard lecturers included former NASA astronaut, Scott Carpenter. The organizers realized the advantage of a cruise approach is maneuverability combined with the use of weather satellite data to position the ship under clear skies. As a result, a small industry was born.[24]

Cruising is a great way to view a total solar eclipse. Here's how the industry works: An eclipse tour group will approach a cruise ship company and charter all or part of a ship. The tour company then resells tickets to this pre-arranged eclipse cruise. Due to this third party charter practice, ships can be completely filled with people whose primary interest is seeing the eclipse, or the passenger population can be mixed and include people who booked the trip for other reasons. Sometimes two or more tour operators will share the same ship, filling it completely with eclipse passengers. All things being equal, choose a cruise where all passengers are sailing because of the big event. There is likely to be a larger group of experienced eclipse chasers onboard and the astronomy program will be richer, making for a more rewarding viewing experience.

Onboard there will be lectures and perhaps films relating to eclipses and astronomy in general. Before the big day, your tour operator will hand out safety glasses, provide instructions, and tell you what time to be out on deck for first contact. Best of all, you don't have to worry about getting to the umbral centerline. The ship's captain will consult with weather experts, and will move the ship to an area with the most likelihood of clear skies. The best part—bathrooms are close, and the buffet is closer. The toughest decision you'll make all day will be whether to eat lunch before or after totality. To find upcoming eclipse cruises, simply search for "eclipse cruise" on the Internet.

Total Solar Eclipse Cruise, February 26, 1998 off the coast of Curacao in the Caribbean Sea.

Viewing an Eclipse from the Air

Again, I recommend viewing the 2017 eclipse from land, however for completeness I will discuss the advantages and disadvantages of viewing from the air. Eclipse chasers will sometimes charter a large jet to chase the eclipse shadow when totality is especially short or when the shadow falls on a geographically or politically inhospitable area of the globe. There is a long history of scientists viewing eclipses from onboard aircraft, however the first commercial eclipse flight took off from Perth Australia in 1974. Half the seats from the left side of the 727 plane were removed so people could set up their gear.

The primary disadvantages of viewing an eclipse from an aircraft is cost. Other minor inconveniences exist such as small windows and less room to position yourself within the aircraft. Also, photographs are taken through windows not designed to be perfectly optically clear and distortion free.

However if you can afford it, the advantages overwhelmingly outweigh the disadvantages: you are far above the clouds and inclement weather, hard to reach locations are not show-stoppers, you have a broad, panoramic view as the shadow traces a path along the ground, and the environment is comfortable with restrooms at hand. Of course the number one advantage is the ability to extend totality. On July 11, 2010, an Airbus 319 (top speed of 541 miles per hour) was used to set a new world's record for the longest eclipse observed by civilians; duration of totality was extended from about 5 minutes on land to 9 minutes and 23 seconds onboard the aircraft.

Extending totality to over one hour: On June 30, 1973, an early version of the Super Sonic Transport (SST) Concord—with a top speed of 1,300 miles per hour—was used by Los Alamos National Laboratory scientists to extend duration of totality to 74 minutes as the shadow passed over Africa. That's more than 10 times the duration possible for observing totality on land.

For the 2001 total eclipse, astronomer Glenn Schneider (Steward Observatory, University of Arizona) was scheduled for a final planning meeting with airline operators of his chartered Concord when Air France flight 4590 crashed in Gonesee, France. This would have been the world's first opportunity for commercial eclipse travelers to experience totality of over one hour. Shortly thereafter, Concord flights were halted permanently.

General Planning Considerations

A Multi-Day Event

The August 21, 2017 Eclipse is on a Monday. For observers on the West Coast, first contact is in the morning between 9 and 10 AM. This probably means sleeping locally the night before.

Why not make a mini-vacation out of the event? Depending upon the distance from your home, plan to take off work for a few days or perhaps a full week.

For example, from Las Vegas, you could start the drive to the centerline on Saturday, spend the night on the road and arrive in time to scope out a viewing sight Sunday afternoon. After the Eclipse on Monday you can drive home, completing the return trip on Tuesday. That's just two weekdays away from the office.

If you have more time, why not spend a few days visiting Yellowstone National Park, which happens to be just a few hours north of the eclipse path?

If you live within a few hours morning drive of the umbral shadow, I still urge you to consider spending the night somewhere close to your viewing site. Traffic may be heavy as you try to drive out the morning of the event. Imagine the time for first contact is approaching, and you are stuck in traffic somewhere outside of the umbra.

Request Time-Off Now

Monday, August 21, 2017 is likely to be the most popular vacation day of the entire year. If you work in a technology company with fellow nerds it may already be too late! Perhaps the boss will just give in and close the place for a few days.

Make Reservations as Early as Possible

Everyone and their mother may be trying to reach that perfect spot you picked to watch the eclipse. After all, the centerline is a line! A long line, but an infinitely narrow line nonetheless. Only a few roads will parallel or intersect the path of totality. Only a limited number of hotels will be a short drive from the umbra. Lodging will fill up fast. If you intend to fly to your location, both airline and car rental reservations need to be made well in advance.

You cannot treat this like any other trip, as demand along the umbra will be fierce. Normally, your personal experience will determine how far in advance to nail-down accommodations; this time I urge you to make reservations much earlier than usual. I suggest trying to book 12 months in advance. Procrastinate, and you may end up sleeping in the back of a car.

Plan a "Micro-Site" and Arrive Early

Don't just drive towards the centerline the day of the eclipse and then pull over to the side of the road. That can be dangerous, particularly as drivers are apt to be distracted by the crowds of people gathering to see the event. Select your micro-site (park, schoolyard, etc.) in advance. Avoid setting up in a parking lot where drivers may roll through. You won't have time to watch for cars as you're staring up at the sky. Be safe.

Why not arrive at the site well before first contact, spread your blanket and enjoy a picnic? If at all possible, hook up with other people, particularly folks hauling telescopes and pro-style camera gear.

Guidance from experienced eclipse chasers can really make a difference; most are more than happy to share their love of the chase.

What to Bring

Plan to bring all your gear from home, as it will most likely be impossible to get certain items like eye protection once you are on the road. If you purchase six months in advance, you should have a wide selection of possible gear. Wait until the month before, and you may find your first choice in eye protection (perhaps those welder's goggles) are no longer available.

Order your camera and binocular filters at least six months before the event. I waited until four months before my last annular eclipse trip, only to find the solar filter for my new telephoto lens was sold out. Also, don't forget to test your equipment beforehand.

Here is a list (in priority order) of what I recommend bringing. You may be surprised that I rank a pair of good binoculars and filters near the top of this list. Experience has taught me that binoculars increase my enjoyment of the eclipse immensely. I hope you will bring a pair and see this for yourself.

1. Eye Protection. Don't show up without at least one of these for each person in your group; and don't plan on sharing them. Most models will fit over a pair of glasses, and you can purchase many for less than ten dollars. Running the risk of someone in your party looking briefly at the sun is not worth it. Bring extra sets if you feel especially generous. You will be surprised at how many people travel all the way to the umbra and don't think to bring a pair. These are also necessary to safely view the partial phase for people outside the umbra.

   I have met people that believe they will be happy watching the partial phase by observing a real-time projection of the eclipse on a piece of white paper. These "pinhole projectors" can be made by simply piercing a small hole in cardboard. A more elaborate version can be made by mounting a pair of binoculars on a tripod or other support.

I believe telling people they should observe in this way is a huge disservice. First, because this can scare people from directly viewing totality with the proper eye protection. Second, when people using pinhole projectors see everyone else at the site looking upward, the devil on their shoulders will come knocking. This is the time to be a hero with that extra pair of safety glasses.

(Note: Chapter 8 of this book, "Safely Viewing the Eclipse," contains further information on eye safety.)

2. Binoculars with Removable Solar Filters. You don't absolutely need binoculars to enjoy a total solar eclipse, but they add a whole lot of fun and awesomeness if you have them. At my first eclipse I was mesmerized when someone was kind enough to let me share theirs. The next eclipse, I had my own.

With the naked eye, the size of the Sun will appear to be the same size as a Full Moon when it is high in the sky (about half a degree of visual angle). Human vision encompasses 180 degrees from left to right so a half of a degree is pretty small. The Moon will appear to be about the size of half a pinky finger when your hand is held at arms-length in front of your face. You won't be able to make out many details.

A pair of binoculars can enlarge the eclipse to fill half or more of your total visual angle. Details become sharp and clear. A modest pair of 7X magnification binoculars will increase the apparent size of the Moon to about the width of two fingers held at arms-length from your face. A pair of 10X binoculars will make the Moon about 3 fingers large and a 25X magnification will make the Moon about the size of your fist.

Falling back on the theater analogy, it's like having a seat in the far rear of the balcony. The binoculars let you see the expression on the actor's faces. If you can afford to bring two pairs, it's even, better as I guarantee your spouse, kids or friend will all want to look (usually during totality).

I recommend 7x35 or 7x50 binoculars. These have a 7x magnification factor and fairly large objective (front) lenses (35mm and 50mm respectively). The larger lenses cost more but they gather significantly more light, making them far easier to look through. Although not an essential feature, you can shop for binoculars that are electronically or optically stabilized as it is difficult to keep your hands steady while searching the Moon's craters for signs of life!

What size binoculars? There are two classes of binoculars to consider: those that are small enough to be hand-held and those requiring a tripod. Higher magnification means more weight and more image jitter. The first number represents magnification and the second indicates the size of the front lenses (in mm). The larger the lens, the better view in low light situations.

Handheld binoculars are simple and comfortable to use when looking at objects high in the sky. They have the added bonus of being versatile—you can use them for many other applications. 7x50 binoculars are a common handheld size. Keep magnification below 10x in order to maintain steadiness while holding by hand.

Larger binoculars with high magnification are made to attach to a tripod or other mount. The reasonably priced Celestron Skymaster series of binoculars are well-known among astronomers. Their 20x70 is a great choice for eclipse viewing (street price $70). You will need a tripod mount for 10X or higher magnification.

The Moon's apparent size in the sky is one-half of a degree (0.5). What does this mean? A full circle is 360 degrees and when looking forward, humans can see about half that or about 180 degrees. When we say the Sun has a "size" of one-half degree, we mean that out of all the 180 degrees humans can see, the Sun appears to take up only half of one of the 180 degrees. That's pretty small.

Since the Moon's apparent size in the sky is one-half degree (0.5), using a pair of 25x70 binoculars with a 2.7 degree field of view, the Sun will fill 19% of the view (0.5/2.7).

Manufacturer	Model	Mag x Lens	Field of View	Size of Moon	Pounds
Nikon	Action Extreme	7 x 50	6.4	8%	2.21
Nikon	Aculon	8 x 42	8	6%	1.69
Celestron	Skymaster	8 x 56	5.8	9%	2.38
Nikon	Aculon	10 x 42	6	8%	1.69
Nikon	Aculon	10 x 50	6.5	8%	2.00
Nikon	Aculon	12 x 50	5.2	10%	2.00
Celestron	Skymaster	15 x 70	4.4	11%	3.00
Celestron	Skymaster	25 x 70	2.7	19%	3.25
Celestron	Skymaster	20-100 x 70	1.25	40%	3.31

Table of sample binocular sizes. Offerings with higher magnification but same "size of Moon" should render finer details.

Binocular *angular field of view*: Binocular specifications indicate their field of view in one of two ways, linear or angular. The angular metric (sometimes called "actual angular") is the angle you can see through the binoculars as measured in degrees. A linear specification such as "420 feet at 1,000 yards" means that if a train was 1,000 yards away from you, you would see 420 feet of it at any one time. Many manufacturers only specify using "linear field of view." To convert to linear to angular, divide the linear number by 52.5. In this example, 420/52.5 = 8 degrees.

I also use my hand-held binoculars at the theater, concerts and sporting events. If you don't already have a good pair, you'll enjoy owning one. The upcoming eclipse is a perfect excuse to splurge.

Do you need solar safety filters for the binoculars? Yes. Without filters, they are not usable during the long partial phase of the eclipse. You and your loved ones will want to look through them beginning with first contact. The partial phase is also a great time to learn how to adjust binocular focus, before the crunch of totality begins.

It is not safe to use your protective eye filter between your face and the binoculars (see sidebar). You must have appropriate filters that are mounted on the Sun side of the binoculars. Note that filters are not usually made by the same manufacturer as the binoculars, so you will need to purchase them aftermarket to fit your pair. Sources of binoculars and appropriate solar filters are covered in Chapter 12, "Resources."

> Solar filters that fit on the Sun side (the large lenses) of the binocular are absolutely necessary. Do not use binoculars during the partial phases without them. Not even for a second. The Sun is being concentrated by the lens and can quickly cause eye damage without the proper filters. DO NOT use your naked eye protection (welder's goggles, or whatever) between the binoculars and your face instead of proper filters fitted to the binocular's Sun side. The unfiltered, focused Sun can cause considerable heat that can crack a glass lens or quickly destroy a plastic substrate filter.

3. Chair(s) and a Picnic Blanket. If possible, bring chairs with reclining backs for obvious reasons. Blankets can be used to "mark off" and reserve your observation area. It seems surprising that this is needed, but I've been to many eclipse events where I thought there would be plenty of room to spread out, only to find trees, poles or billboards in the way, narrowing down the usable real estate. Use the blanket to stake out your territory. On the deck of a cruise ship, people use masking tape to delineate their observation area.

4. Hat, Sunscreen and Lip Protection (Chapstick), insect repellant. You may be out in the sun for a while. Think long pants and a long-sleeved shirt if you are prone to sunburn. When the Sun goes out during totality, you may encounter mosquitoes. It's best to be prepared.

5. Jacket or Sweater. Even if it is warm all day leading up to the eclipse, it will get cool or even cold during totality.

6. Water. Stay hydrated; a cooler with ice is nice. You will be outside and exposed longer than you think. I like to hit the local bar afterward for my beer or champagne, but many people bring them along to the viewing site. Assign a designated driver if necessary.

7. Snacks. At a minimum, one granola bar per person. You'll thank me later.

8. Something to Pee In. Go ahead, snicker out loud if you must. This may not be necessary if your site has facilities nearby, however, you might arrive at that elementary school playground in the middle of nowhere and everything (including the restroom) is locked up. This happened to me once.

Let us suppose you get to the intended site and the bar or restaurant next store gets fed up with providing restroom access to the unusually large crowd. The woods may be an option, but it is also possible there may not be natural camouflage nearby.

Once at the site, develop a restroom plan so everyone knows what to do. You want to be prepared in case someone needs to take care of business 10 minutes before totality. Don't assume that making a pit stop on the way to the site will be sufficient. In my experience, it never is.

9. Smartphone. With a cellular connection, you can check the weather before leaving the hotel on eclipse day. It's also handy for finding directions to an alternate site if you get clouded out. The phone can be used to find a micro-site (such as a baseball field) and its GPS can be used to verify your position on the centerline. See the Chapter 12, "Resources" for the NASA eclipse track overlay website.

10. Camera. I recommend that first timers don't try to photograph the sky during the eclipse, especially during totality. It goes by quickly, and you are better off not being distracted with camera settings, etc.

If you do decide to shoot images or video of the eclipse, you will need a high focal length lens, or the image of the Sun will be rather small. You also require special solar filters to safely capture images on either side of totality. For more information see Chapter 11, "Photographing the Eclipse."

Having said that all this, bring a camera anyway. It's great for capturing the joy on people's faces just after totality ends. Alternatively, point a video camera at the crowd and let it run. You'll find the audio track does a great job of capturing the excitement.

11. Business Cards and a Pen. Bring something you can hand out with your name, email and snail mail address. No business cards? Make a few handmade "personal cards" in advance. Hand these out to the folks with the big camera lens when you ask them to email you a few photos of the eclipse. The pen is for writing down the other person's information because most people won't have thought to bring their own cards.

12. Straw Hat. The Sun becomes an increasingly better point source as totality approaches; shadows of ordinary objects appear sharper. If you hold a straw hat or anything with a mesh of small holes (try a colander) close to the ground, you will see tiny projected images of the crescent Sun. This will work with leaves of trees that overlap, and you can even cross the fingers of both of your hands and observe the Sun's crescent image projected. You may have noticed sharper shadows around sunset at the beach. The phenomenon is the same. As the Sun dips below the horizon, many of its rays are cut-off by the Earth and the remaining rays come to us from a narrow, almost point source.

13. Thermometer. For an easy and fun scientific experiment, bring along an outdoor thermometer and measure the temperature before, during and after totality. There will be a lag in time between the Sun being obscured and the temperature change; you won't see it start to rise again until well after the Sun returns.

14. Eye Patch. This tip is from a very experienced umbraphile but will require some explanation. It takes about 20 minutes for the human eye to adapt to darkness by dilating (opening) the pupils and changing the photochemistry of your eyes to increase their light sensitivity. "Dark adapting" your eye prior to totality will increase your ability to observe the corona. Place an eye patch over your dominant eye about 20 minutes prior to totality and then remove it just after totality starts. This will make a big difference in what you will be able to see in the fine structures of the corona and prominences. The downside is that you will be using only one eye to observe the end of the first partial phase and also the first time the diamond ring appears. To determine which one is your dominant eye, hold your hand out-stretched in front of your face with your thumb sticking up. Take turns looking through one eye at a time. The eye that does not shift when you open and close it is your dominant eye.

Eclipse Observation Etiquette

Everyone at the site traveled a good distance to see this event, so please be courteous to the other people who turned out for the eclipse.

- Never use flash on your camera. This includes when you are taking selfies or candid shots of people. If you don't know how to set your camera's flash to off, please ask. In astronomy circles, an errant flash at the wrong time could be grounds for a lynching.

- Don't play music.

- Please, no smoking. If you need to smoke, make sure you are downwind, and the smoke is moving away from others.

- Try to keep children quietly occupied.

- When in conversation with others, be aware of how loud you are chatting. This is not the time to catch up on gossip. It is especially annoying when everyone is trying to concentrate on the sky, and you and your conversation partner have little interest in what is happening with the eclipse. This is not a little league game where it is perfectly appropriate to catch up with neighbors or relatives. Walk well away from other observers if you find yourself being pulled back into excited conversation.

- Don't walk in front of someone's telescope or camera line-of-sight.

- During totality, don't ask to look through someone's telescope unless you've struck up a particularly friendly conversation and managed to get an invite. If you stand close and look envious, they might welcome you to peek, but I would wait for an invitation.

- Well before or after totality, it's fine to ask politely if you can look through someone's telescope but don't do this when they have a camera attached to the eyepiece unless you just want to look at the camera's LCD display.

- Be aware that even minor vibrations can ruin the setup of a telescope. Don't touch (or heaven forbid, turn a knob on) someone else's scope unless you ask permission first. You may be invited to look through the viewfinder, but you are expected not to touch the barrel, lens or any other part, even with your eye or eyeglasses. If things don't appear right, ask the owner to take a look and let them make adjustments.

- No shade umbrellas, please. You could use one before first contact, but I think it's rude at any time in this context. You will be turning perfectly good observing real estate into unusable space.

- Don't bring your pets. Have you ever watched your dog or cat around a fireworks show? They are not happy campers. Even though the eclipse will be silent, your pet will most likely freak out and start howling and barking. Something is just plain wrong with the world, and they can sense it. Your pet will not only be scared and in distress but will be annoying others around you. It's not going to be fun for them, and you'll end up focused on calming Rover instead of enjoying the wonder. Even for intellectually prepared humans, totality reaches inside and pulls at the strings of a primeval prehistoric fear. Your pets will react to this. Leave Rover at home.

Chapter 7. Choose a Viewing Site:
The Best Places to Go on Eclipse Day

So far this book has answered four questions: What is a total solar eclipse, why should you go see one, how does it happen and what to bring on a trip to view it? Now we will focus on addressing the "where to go see it" question.

In this chapter, you will select a site from which to view the eclipse. Your choice will depend upon what part of the country you live in, or if you're a visitor, where you might consider spending time before or after the event.

We'll start with a discussion of the factors to consider when selecting a site. After that, you will make use of the maps included in the appendix of this book to narrow down options.

Near the end of the chapter, we will discuss what time of day you need to be at the eclipse site.

Things to Consider When Deciding Where to Go

Staying on the Centerline

As mentioned previously, to witness totality on eclipse day, you have to be somewhere under the umbral portion of the Moon's shadow. The umbral shadow will sweep across the country as a dark, approximately round shadow spot about 66 miles in diameter. Think of this shadow as following a path marked by a centerline, with the shadow falling 33 miles on each side of this line (half of 66 miles). Maximum duration of totality will be experienced when you are located on the centerline. On either side of this line, the duration is reduced progressively until you cross the edge of the umbra, at which point you will only see a partial eclipse. Sixty-six miles is the average diameter of the umbra; it actually varies between 62 miles on the west coast and 72 miles on the east coast. As you move away from the centerline, duration fall-off is slow within the first 20 miles, and then accelerates as the following table illustrates:

Distance from Centerline (Miles)	% of Totality Duration
0	100%
10	95%
15	89%
20	80%
25	65%
30	42%

I recommend remaining within 10 miles of centerline, if possible. Observing from up to 20 miles away however, is also reasonable.

Duration Changes across the Shadow Path

Duration of totality also varies throughout the country as you move from west to east along the eclipse path. When the umbra first touches down on the Oregon Coast, duration on the centerline is about 2 minutes. As the shadow sweeps across the country, duration slowly increases until it reaches a maximum of 2 minutes and 40 seconds on the centerline near Carbondale Illinois. After that, it decreases slightly, reaching 2 minutes and 34 seconds at the South Carolina Coast. Unless you plan to fly to your chosen location, this modest change in duration should not be a big factor in your decision. An exception to this may be appropriate if you live in the west as totality is particularly short when the shadow first makes landfall in Oregon. The difference between observing along the coast of Oregon (2 minutes, 0 seconds) compared to Idaho Falls (2 minutes, 19 seconds) is an additional 19 seconds of totality or about 16%.

Note that in the example above, the duration difference of 19 seconds was caused by a longitude (east-west) change of about 600 miles. In contrast, moving away from the centerline, towards the outer edge of the umbral shadow just 17 miles will cause the same drop-off in duration. If possible, it really pays to observe the eclipse from somewhere close to the centerline.

Weather

The three most important factors in real estate are: location, location and location. After making sure you pick a place on or very near the centerline, the three most important factors in choosing where to watching an eclipse are: weather, weather and weather.

Historic atmospheric trends need to weigh heavily in your site selection since a sky full of clouds can quickly render all your careful planning useless. It may well be worth traveling hundreds of extra miles, to reach an area with a better chance of clear skies.

If this sounds pessimistically fatalistic, don't let it deter you. I have never missed the spectacle of totality due to weather. The same goes for my eclipse chasing friend, Craig Small, who has seen over 30 total solar eclipses. It is simply a matter of choosing a location with a decent record of clear skies, and then staying flexible the night before and the morning of the eclipse. Just prior to the eclipse, if the weather at your primary site is forecasted to be bleak, you can always head for a better location.

Here is Jay Anderson's preliminary outlook for the August 21, 2017 eclipse: "The summer thunderstorm season is winding down and retreating southward and the Arizona monsoon is breaking and the storm-carrying jet stream has not yet begin its journey from Canada. The dry and generally sunny fall season is about to begin. After a 38-year eclipse drought, this one arrives to the open arms of a friendly August climatology."[25] Jay's website (listed in the Resources section) is a great stop for climatology information along the umbral path.

Plan Alternatives

After selecting the prime site, think about what would happen if you woke up the day of the eclipse and discovered the weather forecast had turned gloomy. That's why you need second and third alternative locations in mind. Consider the roadways and likely traffic patterns on the day of the eclipse. If necessary, is there a multi-lane highway to get you quickly from your hotel to an alternative location?

The Day before the Eclipse

No matter where you plan to view the eclipse, remain vigilant in the days leading up to August 21, 2017 and be prepared to make last-minute changes. Links to weather forecasting sites are provided in Chapter 12, "Resources."

While you are checking the weather, also look into any wildfires that may be burning nearby. The smoke can obscure your view. This is especially important if you are viewing the eclipse from the dry Pacific or Mountain states.

Things may look good on a map but also check with highway and municipal road authorities (over the Internet) for scheduled road closures. Getting stuck in a slow moving detour because they're repairing the only bridge for 50 miles around would not be fun.

The Five Steps to Select Your Viewing Site

Follow these five steps to zero-in on your site:

1. Get acquainted with the path of totality on the U.S. Map.
2. Review cloud cover history across the shadow path by studying the weather charts in this book.
3. Select your region and find a town from the maps in this book.
4. Use a satellite view on the Internet to check out terrain and roads. Choose one or more micro-sites for eclipse day, such as a park, a schoolyard, etc.

5. Lookup the eclipse start time for your location by finding the *local circumstances table* containing first through fourth contact times.

The next few pages consider each of these five steps in more detail.

Step 1. Get Acquainted with the Path of Totality

Review the path of totality overlaid on a map of the U.S., taking note of any special places that peak your interest. Perhaps a relative lives somewhere not too far from the path, and you are due for a visit. If you have always wanted to see Yellowstone National Park, that might influence you. Consider the distance you are comfortable driving. If you're willing to fly, look for locations with easy access to cities served by major airlines.

To get started, here is a list of select cities and towns that are somewhat close (but not on) the path of totality in 2017: Bend, Portland, Boise, Jackson (near Yellowstone), Cheyenne, Salt Lake City, Denver, Omaha, Topeka, Kansas City, Saint Louis, Louisville, Evansville, Chattanooga, Knoxville, Gatlinburg, Des Moines, Atlanta, Augusta, Charlotte, Spartanburg and Myrtle Beach.

Map of the umbral path across the U.S. Maximum totality is 2 minutes, 40 seconds near Carbondale, Illinois. Observers inside the narrow red lines will see a total eclipse. Between the red lines and the first set of black lines, the eclipse will be partial with between 99% and 80% of the Sun covered. People within the next band of black lines will see between 79% and 60% of the Sun covered.

Step 2. Review Historic Cloud Cover

Familiarize yourself with the places of lowest cloud cover along the eclipse path. Experienced umbraphiles (eclipse chasers) will cluster around the few locations with the best weather prospects. This is why if you've selected properly, you are likely to run into veteran eclipse chasers on the big day. Poll a group of 100 people, and you might discover 90% have selected the same two or three areas from which to observe. Some people may end up in a school yard and some may view from a baseball field, but they all chose the same town along the centerline with the most promising weather outlook.

The two primary sources of historic cloud data are ground reporting stations (usually airports) and satellite imagery. Do not compare ground reported percentages with satellite reported percentages as the methodologies are different. Also, only use cloud cover information to compare one site with another and not to decide the absolute probability of seeing the eclipse.

With ground reported data, we only have information for scattered locations along the path. Be careful not to assume an area 50 miles away has a similar cloud history to the reporting station. For example, if there are mountains to the west of the station, rising moist air carried from the Pacific will create clouds on the windward (west) side, even though east of the mountain reports little cloud activity. In most cases, the mountain acts as a cloud barrier, making the east particularly dry.

Weather information is provided for informational purposes only. Selection of viewing location is the responsibility of the reader. The Author and Publisher of this book are not responsible for any individual's failure to successfully view the eclipse due to weather, logistics, or any other cause.

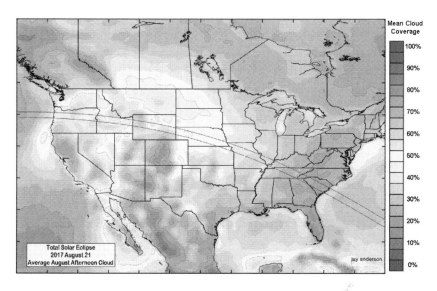

Average August afternoon cloudiness. Derived from 22 years of satellite observations. Best prospects are Northern Oregon, Idaho, Central Wyoming and Western Nebraska. Courtesy of Jay Anderson, www.eclipser.ca

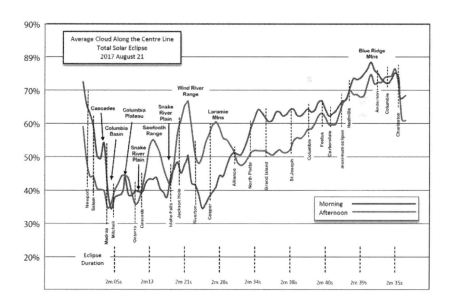

August daily cloud cover along centerline derived from 20 years of satellite imagery. Cities and towns along the eclipse path are indicated by dashed vertical lines above their names. Courtesy of Jay Anderson, www.eclipser.ca

85

Averages can be misleading. The following table presents a more thorough view of cloud history near the umbral path. Note that the ground reporting stations in the table below are only near the central shadow and may be located outside the umbra. Portions of the cloud cover data from this chart are repeated on the detailed maps. Each region is presented in its own table.

Column 1-3: State, reporting site, community.

Column 4: Percent of days during the month with cloud cover (either broken clouds or overcast). The list is sorted with the best (lowest cloud cover) at the top.

Column 5-9: Percent of days in the month with cloud type indicated.

State	Station	Community	Cloud Cover	Clear	Few	Scat	Broken	Overcast
OR	Ontario Muni	Ontario	16%	77%	7%	0%	4%	12%
OR	Roberts Field	Redmond	24%	50%	18%	7%	16%	8%
ID	Boise	Boise	25%	50%	19%	6%	15%	10%
ID	Pocatello	Pocatello	28%	44%	21%	8%	18%	10%
OR	Baker Muni	Baker City	30%	38%	30%	2%	17%	13%
WY	Riverton Regional	Riverton	30%	30%	41%	0%	22%	8%
ID	Idaho Falls Regional	Idaho Falls	31%	43%	25%	1%	18%	13%
WY	Jackson Hole Airport	Jackson Hole, Teton Village	32%	26%	43%	0%	22%	10%
WY	Hunt Field	Lander, Boulder Flats	36%	22%	32%	10%	23%	13%
WY	Natrona County	Casper	38%	26%	27%	10%	20%	18%
OR	McNary Field	Salem	45%	38%	14%	4%	13%	32%
OR	Aurora State	Aurora	48%	38%	12%	2%	17%	31%
OR	Corvallis Muni	Corvallis	56%	28%	16%	1%	22%	35%
OR	Newport Municipal	Newport	51%	18%	16%	0%	18%	33%

Pacific and Mountain region average August cloud types near umbra, from ground reporting stations. Source: Jay Anderson, www.eclipser.ca

State	Station	Community	Cloud Cover	Clear	Few	Scat	Broken	Overcast
NE	Alliance Muni	Alliance	35%	33%	33%	0%	23%	12%
NE	W. Nebraska Regional	Scottsbluff, Minatare	40%	26%	23%	12%	25%	16%
NE	Kearney Regional AP	Kearney, Gibbon	44%	28%	28%	0%	13%	32%
NE	Central Nebraska AP	Grand Island	46%	26%	15%	14%	23%	23%
NE	Hooker County AP	Mullen	46%	22%	32%	0%	20%	26%
MO	Kansas City Int'l	Kansas City, Platte	47%	22%	19%	12%	26%	21%
NE	Broken Bow Muni	Broken Bow	47%	20%	32%	0%	28%	19%
NE	Hastings	Hastings	48%	21%	32%	0%	32%	15%
MO	Spirit of St. Louis AP	Saint Louis	49%	15%	32%	4%	29%	20%
MO	Cape Girardeau Muni	Cape Girardeau, Jackson	50%	11%	35%	5%	31%	19%
Il	Southern Illinois	De Soto, Murphysboro	49%	12%	38%	0%	34%	16%
NE	Brenner Field AP	Falls City	51%	21%	28%	0%	36%	15%
MO	Columbia Regional	Columbia, Ashland	51%	13%	24%	11%	24%	27%
NE	Lincoln Muni	Lincoln	52%	22%	19%	8%	24%	28%
MO	Jefferson City Mem.	Jefferson City	52%	15%	33%	0%	31%	21%
NE	Beatrice Muni	Beatrice	52%	16%	31%	0%	33%	19%
MO	Rosecrans Mem. AP	Saint Joseph	54%	20%	23%	3%	30%	24%
MO	Saint Louis Int'l	Saint Louis	54%	11%	21%	14%	31%	23%
Il	Williamson Co. Reg.	Marion, Herrin, Carterville	59%	6%	36%	0%	41%	18%
Il	St Louis DT AP	Saint Louis	60%	8%	32%	0%	41%	19%

Central region average August cloud types near umbra, from ground reporting stations. Courtesy of Jay Anderson, www.eclipser.ca

State	Station	Community	Cloud Cover	Clear	Few	Scat	Broken	Overcast
TN	Chattanooga Metro Airport*	Chattanooga	31%	8%	24%	37%	25%	6%
SC	Anderson Regional Airport	Anderson	50%	5%	45%	0%	34%	16%
TN	Nashville International Airport*	Nashville	52%	5%	19%	24%	33%	19%
TN	Smyrna Airport*	Smyrna	52%	1%	47%	0%	40%	12%
KY	Barkley Regional Airport	Paducah	54%	8%	21%	17%	35%	19%
GA	Athens-Ben Epps Airport*	Athens	55%	3%	22%	20%	37%	18%
SC	Columbia Metropolitan	Columbia	57%	3%	21%	19%	35%	23%
NC	Ashville Muni*	Ashville	58%	4%	23%	15%	33%	25%
SC	Greenville Spartanburg	Greenville	58%	4%	20%	18%	37%	21%
KY	Bowling Green/Warren County	Bowling Green	60%	6%	34%	0%	40%	20%
SC	Charleston AFB/Intl	Charleston	68%	3%	14%	15%	39%	29%

Eastern region average August cloud types near umbra, from ground reporting stations. Courtesy of Jay Anderson, www.eclipser.ca

Step 3. Select a Region and Find a Town on the Detailed Maps

Near the end of this book you will find a set of maps. Select from among the three regions that most interest you. The regions roughly correspond to the U.S. time zones: Pacific and Mountain regions are grouped together, followed by the Central and Eastern Time regions.

Following the regional map page, you will find detailed maps showing small towns and highways. Most states are spread across multiple pages to make them easier to read. From the detail maps, you can work your way eastward by flipping to the next page.

You will also find historical cloud cover data for select locations on the detailed maps. This is presented as shaded areas of red, blue and green along with the cloud cover percent. For example, "45% cc" means that during the month of August, the location has seen either overcast or broken skies on 45% of the days in the month. Look for areas marked in green, as these have less than 40% cloud cover probability.

Red: more than 50% clouds

Blue: between 40 and 50% clouds

Green: less than 40% clouds

Locations Look Most Promising

Oregon, Idaho and Wyoming have several locations with better than average histories of very low cloud cover. The area around Madras, Oregon, just east of the Cascade Mountains in the Columbia Basin looks promising. Other prime choices include the towns around Baker City and Ontario in the Snake River Plain north of Boise, Idaho. Riverton, Wyoming (west of Casper) stands out as an attractive spot, partly because of its proximity to Yellowstone National Park and the Tetons. The valley around Jackson Hole near the airport might make a good observing site. Moving east from the Wyoming border, Alliance, Nebraska has particularly good prospects. If you plan on viewing from anywhere east of central Nebraska (including the entire Eastern

Region), the locations here report similar cloud frequency. The best approach is to stay flexible, moving to wherever skies are clear on eclipse day.

Step 4. Use the Internet to Check-out Terrain, Roads and Find Micro-sites

Next you will need to go online to perform further research on the selected area. Use a web browser and navigate to:

www.eclipse.gsfc.nasa.gov/SEmono/TSE2017/TSE2017.html

This is NASA's web page for the 2017 Eclipse. About two-thirds of the way down, you will find a link to the "Interactive Map of the Path of Totality." Follow that link and zoom in to see close-up maps of your selected location.

Turn on satellite imagery (or use Google Earth) and get a really good look at the terrain. Are roads leading to the centerline paved? Identify potential micro-sites by looking for developed public land where you can set up on the day of the eclipse. Good bets are school yards, landscaped parks, public squares, sporting fields, festival grounds, or perhaps a farmer's field. All these will need to be unfenced, or you must be sure there is a way to gain legal access. You don't want to be run off the property as totality approaches! You might want to consider access to restroom facilities, either natural or man-made as your comfort level dictates.

Step 5. Look-up Eclipse Time for Your Location—Get Your *Local Circumstances* Table

What time should you show up? It will take 65 minutes for the umbral shadow to sweep between the Oregon and South Carolina coasts, crossing four time zones along the way. The time that the shadow reaches your chosen site depends on where along the path you are located; the time for first contact is specific to your location.

As discussed earlier, duration of totality is different based on longitude and distance from the centerline, therefore the timing of second and third contact are also unique to your observing location.

While this book provides timetables for a handful of towns, we elected not to include page after page of tables covering hundreds of potential observing locations. Fortunately, there are websites where you can enter your location on a map or type in an address and get a *local circumstances table*. These tables will include the time for each contact event, along with the Sun's elevation and azimuth during the eclipse.

There are also smartphone applications that perform a similar function using your current GPS location or a manually entered address. To find one, search for "solar eclipse time" in your app store of choice. I've downloaded several and they all work about the same.

Fun to think about: The 2017 eclipse takes about an hour to move from west to east across the four time zones of the U.S. (Pacific, Mountain, Central and Eastern). People on the east coast will see the eclipse one hour later than people on the west coast, but eastern clocks report it as four hours later. The west coast sees totality around 10:15 in the morning when the Sun is not quite halfway to its noon zenith (elevation= 39 degrees). The east coast sees totality about 2:40 in the afternoon with the Sun a bit off its zenith, at 62 degrees elevation.

Using the online map on the NASA site is one of the best approaches to find a local circumstances table. Find the website here:

http://eclipse.gsfc.nasa.gov/SEgoogle/SEgoogle2001/SE2017Aug21T google.html

After opening the site, click your location of interest on the map. A table similar to the one shown below will pop open. This example is for Madras, Oregon, almost directly on the centerline. Time in this table is expressed as UT (Universal Time). We will discuss how to convert to the applicable local timezone in the next section.

Visit Chapter 12, "Resources," for addresses of other websites where you can find a local circumstances table for your chosen viewing site.

Lat.: 44.6348° N
Long.: 121.1301° W

Clear Marker

Total Solar Eclipse
Duration of Totality: 2m02.0s
Magnitude: 1.011
Obscuration: 100.00%

Event	Date	Time (UT)	Alt	Azi
Start of partial eclipse (C1) :	2017/08/21	16:06:43.6	29.4°	102.9°
Start of total eclipse (C2) :	2017/08/21	17:19:36.2	41.5°	119.1°
Maximum eclipse :	2017/08/21	17:20:37.1	41.6°	119.4°
End of total eclipse (C3) :	2017/08/21	17:21:38.2	41.8°	119.6°
End of partial eclipse (C4) :	2017/08/21	18:41:06.0	52.4°	143.8°

Table of local circumstances for Madras, Oregon, 21 August, 2017 Source: NASA website map.
http://eclipse.gsfc.nasa.gov/SEgoogle/SEgoogle2001/SE2017Aug21Tgoogle.html

In the table above, C1 is first contact, which is the start of the eclipse. You should plan to be settled at your site a comfortable 30 minutes or more before this time. C2 (second contact) is the start of totality. Maximum eclipse is halfway between second and third contact (when the Sun and Moon are centered). People begin to pack up about 20 minutes after C3 (third contact), although some folks remain observing until C4 (fourth contact).

Universal Time Conversion

You will notice that the times given in these local circumstances tables are frequently expressed as UT. What the heck is a UT? Universal Time is what astronomers use. For our purposes, the terms UT, UTC and GMT all refer to the same thing. The breakout box explains the differences, if you are so inclined.

In the winter, all the U.S. states are on Standard Time. In the summer, all states except Arizona and Hawaii switch to Daylight Savings Time (DST), so we can work and play outside longer into the evening. In 2017, DST will start Sunday, March 12 and will end Sunday, November 5. Arizona (in the Mountain Time Zone) does not observe DST because with the intense heat, there is no interest in having the Sun hanging around later in the evening. The Navaho Nation is Arizona's exception and does observe DST.

To convert UT to local time, here are two handy tables. Since the eclipse will be in August, use the Daylight Savings Time table (NOT Standard Time). Both summer and winter tables are provided for completeness.

Use in summer when U.S. is on Daylight Savings Time (except Arizona):

Pacific Daylight Time (PDT) = UTC - 7 hours

Mountain Daylight Time (MDT) = UTC - 6 hours

Central Daylight Time (CDT) = UTC - 5 hours

Eastern Daylight Time (EDT) = UTC - 4 hours

Use in winter when U.S. is on Standard Time:

Pacific Standard Time (PST) = UTC - 8 hours

Mountain Standard Time (MST) = UTC - 7 hours

Central Standard Time (CST) = UTC - 6 hours

Eastern Standard Time (EST) = UTC - 5 hours

Therefore, if you're planning to observe from Madras Oregon (Pacific time zone), the eclipse will start (C1) at 9:07 AM (16:07 - 7 hours). Totality (C2) will begin at 10:19 AM (17:19 - 7 hours).

If you are not sure what time zone your target location is in, be careful with the conversion. The following example illustrates the danger: You ask Google, "What time zone is Madras, Oregon in?"

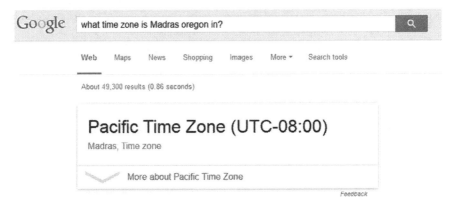

Notice that Google returns UTC - 8 (same as UT - 8) even though this inquiry was performed in July. It is giving us the offset from UTC as if we were operating on Standard Time. Since we will be using Daylight Savings Time on eclipse day, you need to convert to Standard Time (UTC - 8) and then add back one hour to adjust for Daylight Savings Time (C1: 16:06 UT - Google's 8 hours = 8:06 + 1 = 9:06 AM).

Two sets of local circumstances tables are provided below. The first set includes some of the largest major metropolitan areas in the United States. The second set covers select locations along the path of totality. If a location does not fall under the umbral shadow, totality will not be visible, and both C2 and C3 will be missing from the table. All times are local, therefore no conversion is necessary.

You can use these tables to double check conversions you perform on your own.

The following local circumstances table covers select metropolitan areas across the US. It is sorted alphabetically. Eclipse predictions and eclipse track map coordinates courtesy of Fred Espenak, NASA/Goddard Space Flight Center.

State	City/Town	C 1	C 2	C 3	C 4	%	Duration	Elev	Azim
AL	Birmingham	12:00:44 PM			2:58:22 PM	93.5%		66	206
AK	Anchorage	8:21:37 AM			10:13:50 AM	55.6%		19	100
AZ	Phoenix	9:13:53 AM			12:00:26 PM	70.0%		56	122
AR	Little Rock	11:47:06 AM			2:46:31 PM	90.5%		67	184
CA	Los Angeles	9:05:43 AM			11:44:50 AM	69.4%		48	113
CA	San Francisco	9:01:31 AM			11:37:08 AM	80.2%		43	111
CO	Denver	10:23:22 AM			1:14:43 PM	93.3%		57	144
CT	Hartford	1:25:25 PM			3:59:34 PM	73.2%		51	227
DC	Washington	1:17:59 PM			4:01:41 PM	84.0%		56	223
FL	Miami	1:26:56 PM			4:20:48 PM	82.3%		64	242
FL	Orlando	1:19:25 PM			4:14:55 PM	87.0%		64	233
GA	Atlanta	1:05:51 PM			4:01:54 PM	97.1%		65	214
ID	Boise	10:10:34 AM			12:50:07 PM	99.4%		46	126
Il	Chicago	11:54:18 AM			2:42:40 PM	88.9%		59	193
IN	Indianapolis	11:57:53 AM			2:48:39 PM	92.7%		61	199
IA	Des Moines	11:42:50 AM			2:33:58 PM	95.4%		60	176
KS	Kansas City	11:41:12 AM	1:08:33 PM	1:08:59 PM	2:35:54 PM	103.1%	0:00:26	63	173
KY	Louisville	12:59:23 PM			3:51:57 PM	96.1%		62	202
LA	New Orleans	11:57:42 AM			2:57:20 PM	80.0%		71	200
MD	Baltimore	1:18:32 PM			4:01:18 PM	83.1%		56	223
MA	Boston	1:28:27 PM			3:59:29 PM	70.2%		50	229
MI	Detroit	1:03:28 PM			3:47:50 PM	83.1%		57	204
MN	Minneapolis	11:43:56 AM			2:29:04 PM	86.1%		57	176

State	City/Town	C 1	C 2	C 3	C 4	%	Duration	Elev	Azim
MO	Saint Louis	11:49:55 AM			2:44:17 PM	100.0%		63	188
NE	Omaha	11:38:27 AM			2:30:19 PM	98.3%		60	196
NV	Las Vegas	9:09:08 AM			11:53:00 AM	77.1%		51	120
NJ	Newark	1:22:53 PM			4:00:34 PM	77.1%		53	226
NY	New York	1:23:15 PM			4:00:45 PM	76.9%		53	226
NC	Charlotte	1:12:21 PM			4:04:22 PM	97.8%		61	221
OH	Cleveland	1:06:31 PM			3:51:07 PM	83.6%		57	209
OK	Oklahoma City	11:37:07 AM			2:34:53 PM	87.2%		66	164
OR	Portland	9:06:20 AM			11:38:35 AM	99.1%		40	118
PA	Philadelphia	1:21:14 PM			4:01:19 PM	79.9%		54	225
RI	Providence	1:28:05 PM			4:00:28 PM	71.7%		50	229
SC	Greenville	1:09:13 PM	2:38:02 PM	2:40:13 PM	4:02:56 PM	103.0%	0:02:11	63	217
TN	Nashville	12:58:30 PM	2:27:26 PM	2:29:21 PM	3:54:02 PM	103.1%	0:01:55	64	202
TX	Dallas	11:40:23 AM			2:39:21 PM	80.0%		69	166
TX	Houston	11:46:40 AM			2:45:45 PM	73.0%		72	174
UT	Salt Lake City	10:13:58 AM			12:59:36 PM	92.5%		51	130
VA	Virginia Beach	1:21:47 PM			4:06:57 PM	88.0%		56	229
WA	Seattle	9:08:45 AM			11:39:00 AM	93.1%		39	120
WI	Milwaukee	11:53:38 AM			2:40:09 PM	86.2%		58	191
WY	Cheyenne	10:23:52 AM			1:13:52 PM	97.0%		56	145

The following local circumstances table covers select towns near the eclipse centerline. It is sorted west to east. Eclipse predictions and eclipse track map coordinates courtesy of Fred Espenak, NASA/Goddard Space Flight Center.

State	City/Town	C 1	C 2	C 3	C 4	%	Duration	Elev	Azim
OR	Lincoln City	9:04:41 AM	10:16:06 AM	10:18:02 AM	11:36:11 AM	102.6%	0:01:56	39	116
OR	Corvallis	9:04:54 AM	10:16:56 AM	10:18:36 AM	11:37:24 AM	102.7%	0:01:40	40	116
OR	Madras	9:06:43 AM	10:19:36 AM	10:21:38 AM	11:41:05 AM	102.7%	0:02:02	42	119
ID	Idaho Falls	10:15:12 AM	11:33:01 AM	11:34:51 AM	12:58:04 PM	102.9%	0:01:50	50	132
ID	Driggs	10:16:28 AM	11:34:22 AM	11:36:40 AM	12:59:42 PM	102.9%	0:02:18	50	134
WY	Riverton	10:19:33 AM	11:39:03 AM	11:41:17 AM	1:05:20 PM	102.9%	0:02:14	53	139
WY	Casper	10:22:18 AM	11:42:40 AM	11:45:06 AM	1:09:26 PM	102.9%	0:02:26	54	143
WY	Evansville	10:22:23 AM	11:42:46 AM	11:45:11 AM	1:09:31 PM	102.9%	0:02:25	54	143
NE	Alliance	11:27:08 AM	12:49:14 PM	12:51:44 PM	2:16:44 PM	103.0%	0:02:30	57	151
NE	North Platte	11:30:17 AM	12:54:06 PM	12:55:48 PM	2:21:47 PM	103.0%	0:01:42	59	145
NE	Grand Island	11:34:20 AM	12:58:34 PM	1:01:09 PM	2:26:36 PM	103.0%	0:02:35	60	162
MO	Saint Joseph	11:40:39 AM	1:06:28 PM	1:09:06 PM	2:34:38 PM	103.0%	0:02:38	62	172
MO	Jefferson City	11:46:07 AM	1:13:10 PM	1:15:38 PM	2:41:07 PM	103.1%	0:02:28	63	181
Il	Chester	11:51:06 AM	1:18:38 PM	1:21:17 PM	2:46:10 PM	103.1%	0:02:39	64	190
Il	Carbondale	11:52:26 AM	1:20:06 PM	1:22:43 PM	2:47:28 PM	103.1%	0:02:37	64	192
KY	Hopkinsville	12:56:33 PM	2:24:42 PM	2:27:22 PM	3:51:44 PM	103.1%	0:02:40	64	199
TN	Crossville	1:02:28 PM	2:30:59 PM	2:33:32 PM	3:57:08 PM	103.0%	0:02:33	63	208
SC	Columbia	1:13:06 PM	2:41:49 PM	2:44:19 PM	4:06:19 PM	103.0%	0:02:30	62	222
SC	Charleston	1:16:59 PM	2:46:28 PM	2:47:56 PM	4:10:02 PM	103.0%	0:01:28	62	228

Correcting time for the *lunar limb*: The *lunar limb* (or *lunar profile*) is the outer edge of the visible surface of the Moon as viewed from Earth. Baily beads and the diamond ring effect are created as sunlight streams through the mountains and valleys along the irregular edge towards the viewer.

Precise eclipse times for first contact, duration and other time points ultimately depend on when the apparent edges of the Moon and Sun meet. But the irregularity of the lunar limb means the Moon is not a perfect circle. Most local circumstance calculations assume a simple circular lunar radius, halfway between the mountain peaks and valleys.

Event times based on the simple circular Moon model can be off by as much as three seconds, however this approximation is fine to use on the day of eclipse. Occasionally you will find tables with adjustments made for the shape of the lunar limb. The model used for the times stated in this book are not adjusted for the shape of the lunar limb.

UT (*Universal Time*) is based on the Earth's rotation and measured against objects in space. UTC (*Coordinated Universal Time*) is the system we use on Earth to coordinate our daily activities. Our smartphone clocks and our GPS system provide UTC time. There is a process in place to periodically synch up UTC with UT, ensuring they are never more than 0.9 seconds apart. GMT (*Greenwich Mean Time*) is what we called our system before the name was changed to UTC.

Chapter 8. Safely Viewing the Eclipse:
Eye Protection, Lens Filters and How to Use Them

When I was small, my father told me never to touch the parking brake or the transmission shifter in the car. Perhaps you were told never to touch anything on the stovetop. Was this good advice? You bet, we were kids. But think of going through life missing out on the pleasure of driving or cooking: that would be absurd. The advice to "never look" at an eclipse is similarly absurd (for an adult anyway). You just need a bit more information to do it safely. It's like using a pot holder when cooking so you don't get burned.

No Death Rays

Let's get this out of the way first. There is no such thing as special "eclipse rays" that can damage your eyes. It is safe to let your dog outside during an eclipse. He won't get zapped. But, is the sun safe to look at throughout the entire eclipse? Absolutely not. The Sun is a ball-shaped, hydrogen-burning nuclear-reactor that is 856,000 miles in diameter, with a temperature of over one million degrees. It can melt your socks off. Some respect is required here.

Looking at an eclipse is no more dangerous than looking at the Sun. Which is to say, it is just as dangerous as looking at the Sun. You should never look directly at the Sun without proper eye protection, even for a second.

What makes the eclipse dangerous of course, is that normally we are not interested in looking at the Sun. But when all of a sudden something unusual is happening up there, we all want a peek; especially kids and people who may not have access to public warnings. For example, we know that when a solar eclipse takes place over populated areas of third world countries we receive dozens of cases of eye damage.

At a certain age, telling someone to not look at the Sun is like asking them to not think about pink elephants. We would be better off educating them about how to look safely at the Sun by explaining what constitutes proper eye protection.

> "Why did the kids put beans in their ears? No one can hear with beans in their ears. Why did the kids put beans in their ears? They did it 'cause we said no."
>
> The song "Never Say No" from the off-Broadway musical, "The Fantasticks."[26]

Caution Is Warranted

The spectrum of the Sun's rays that reach the Earth include visible light, infrared (IR) and ultraviolet (UV) radiation. The Sun can cause two primary types of damage to our eyes. First, exposure of our retina to intense visible light or UV radiation can damage the ability of our rod and cone cells to respond to visual stimuli. And second, if the exposure is long enough, the near-infrared radiation can cause heating that results in severe thermal injury (burning), destroying rod and cone cells.

This damage can result in total blindness or the creation of a localized blind area in our vision. Our retina does not contain pain receptors so the damage will happen well before we become aware of it. It may even be several hours before our sight becomes impaired. Once damage has occurred, hours later (like a sunburn) your eyes will start to water and feel incredibly painful.

It does not take long for damage to occur (less than one second under certain circumstances). Viewing the Sun through a lens (either a camera, telescope or binoculars) can result in immediate retinal injury because the image is magnified, and the Sun's rays are concentrated in a focused spot.

We all have stories: A friend was replacing the UV light bulbs that sterilized the water in his backyard koi pond. The new bulbs didn't appear to be "on" so he looked inside the tube that was holding them, waiting for visible light. After about 30 seconds, he figured that these bulbs probably don't give off visible light, just radiation in the UV band. He finished the repair. About four hours later his eyes began to feel like sand paper and six hours later he could no longer see. Next stop was the emergency room. Some strong pain killers followed. Two days later his sight returned. Within a week, the pain had pretty much gone away. He was lucky and seems to have side-stepped permanent damage.

When Can I Look Without a Filter?

I want this to be perfectly clear: You should view the *total phase* of the total solar eclipse without eye protection as long as you understand what you are doing. It isn't difficult to do it right; advanced training is not required. It's not like BASE jumping in Yosemite or flying a sailplane. You just need to pay attention to a few simple rules: totality, eye protection off, all other times, eye protection on. You can touch the pot on the stove with your bare hands, just not when it's hot!

The only time it is safe to view the Sun directly or through any lens without proper protection, is during totality (assuming you are observing from somewhere within the umbral shadow). The start of this period (when it's safe to look at the Sun unprotected) is after seeing Baily beads, the diamond ring and second contact and the Sun is completely covered by the Moon. I said *after* all these events. Once these things have occurred, the disk of the Moon completely covers the bright portion of the sun and filters can come off. If you are with a group of experienced eclipse viewers someone may announce out loud, "glasses off," "filters off" or something like that. That said, make sure *you* understand when it's safe to look without safety glasses or lens filters; don't assume that the person calling out "glasses off" knows what they are doing.

So when do you put the eye protection back on? This can be a bit tricky. Although totality might be forecasted to last 2 minutes and 10 seconds, don't rely on counting the seconds as timing could be off for various reasons. You will need to start using eye protection again just before the first sliver of Sun appears. By closely watching the edge of the Moon you can tell when the disk is just about to start revealing the light of the Sun. Just before the end of totality, stop using your lenses (binoculars, etc.) and just look without any magnification. If you move your eye protection back in position with a slight delay, not peering through a lens will be gentler on the eyes. It is important to remember to slap the glasses back on before you see the first bit of Sun from behind the Moon. Don't wait for the diamond ring to form.

> Remember, the exposed crescent Sun can be just as damaging to your eyes as the full Sun. As the eclipse progresses, temptation will build and curiosity can drive people to look at the thin crescent without proper protection. Don't let this happen to you or your loved ones.

Unsafe Eye Protection- DO NOT USE THESE!

I know plenty of people who don't bother traveling to the umbra and then think they can step outside and sneak a peek. They may get away with it (without permanent damage) but if they had spent a few minutes a week earlier and shelled out a few dollars they would be prepared. I also know people (so do you) that think stacking three or four pairs of sunglasses is going to protect them. It won't. For this reason, I include a list of things that are not proper eye protection, in addition to what is safe to use.

The following is a short, partial list of things NOT to use or do. Most of these either don't attenuate (reduce) enough visible light or don't attenuate the infrared band at all:

Bad scene #1: You plan on looking only during totality. Totality is so short, and the partial phase before totality is so long, you will be tempted to look up out of boredom and curiosity.

Bad scene #2: A stack of three or four pairs of sunglasses.

Bad scene #3: Smoked glass.

Bad scene #4: Photographic neutral density (ND) filters.

Bad scene #5: Polarizing filters.

Bad scene #6: "Space blankets."

Bad scene #7: A strip of black and white or color negative film. This works only with certain types of black and white negatives, and only if they don't have any image on them. They need to be completely exposed and developed so that the silver halide covers all of the film. This is hard to make in today's digital world, so why bother?

Bad scene #8: CD or DVD disks. Bad idea even if the hole is covered. Same for blue-ray disks in case you were wondering.

Bad scene #9: Floppy disks. (What the heck is a floppy disk anyway?)

Safe Eye Protection

Proper eye protection will transmit less than 0.0032% of visible light. When you view a typical outdoor scene through proper filters, things will appear nearly black. Very little light should be coming through. These are all much darker than any pair of sunglasses you've ever used. It is almost like looking through a piece of black paper. Proper eye protection will also pass only a limited amount of infrared light, usually less than 0.5%. These filters should also attenuate UV light. Proper filters will be quality controlled, having no coating defect of more than 0.2 mm and no more than a single defect in any 5 mm area (or some similar specification). The best way to ensure you get proper eye protection is to purchase it at a telescope shop or order them online, making sure they are designed for solar (or eclipse) viewing. Look for glasses that are certified to ISO 12312-2:2015. This means they have been specifically designed for solar viewing. Vendors and websites are listed in Chapter 12, "Resources."

Safe to use #1: Black polymer. Carbon particles suspended in a resin film. Produces a pleasing yellow image of the Sun. Generally the least expensive filters. Available in cardboard glasses, plastic molded frame glasses and sheets for do-it-yourself types and to make camera or binocular filters. Easy to find ISO certified. I've heard reports from people that sometimes the pre-made cardboard glasses don't fit perfect if you have eyes that are unusually close together. This can especially be a problem for children and teens. The best idea is to check the comments about fit on particular supplier websites.

Observing with cardboard solar "glasses." March 29, 2006,
off the coast of Turkey and Greece.

Safe to use #2: Aluminized polyester film and cardboard glasses made for solar observation. This is sometimes erroneously referred to as "Mylar." Mylar is a trademarked product of Dupont and they don't officially make the material for use as solar eye protection. It's made from plastic film with a thin, smooth surface deposit of metal to block the Sun's radiation. The Sun's image through these filters is usually blue-white. Several suppliers make aluminized polyester film with the quality controlled defect rate necessary for eclipse viewing. You can buy this in sheets or already mounted into inexpensive cardboard "glasses." Note that the aluminized polyester film you can buy at the hardware store for gardening purposes is not suitable for eclipse viewing. Purchase your aluminum polyester film or glasses from a reputable manufacturer that has designed it for this purpose. See Chapter 12, "Resources" for suppliers.

Safe to use #3: AstroSolar™ Safety Film. Manufactured by Baader in Germany. Users report excellent photographic results with this material. The image is uncolored. You may have seen this material as a thin crinkled mirror-silver film. Installation with some wrinkles is normal and acceptable. Three versions are available: 1) Astrosolar™ Safety Film (OD 5.0) for safe solar observation with magnification and photography; 2) AstroSolar™ Silber Folie (OD 5.0) for use in low magnification applications (such as eyewear); and 3) AstroSolar™ Foto Film (OD 3.8), which is not to be used as eye protection and is too fast for still photography. It is only used with astronomical telescopes or certain video applications by persons experienced with solar photography.

Safe to use #4: Number 14 welder's goggles. I have used these myself on occasion. The glass is mounted as a pair of goggles with a head strap, so they stay on your face. Although they have been used by eclipse viewers for years, these appear lower in my list because they are not ISO certified for solar viewing.

Observing with #14 welder's goggles. March 29, 2006, off the coast of Turkey and Greece.

Safe to use #5: R-G solar film from Thousand Oaks Optical.

Safe to use #6: Optically coated glass made for solar viewing. Be sure your choice is made for direct viewing, not just for photography.

Camera, Telescope and Binocular Filters

Filters and protective lenses that are intended to be placed directly in front of your eyes are not appropriate (and can be dangerous) to use between your eyes and a device with a lens. When using cameras, binoculars and telescopes for viewing the sun, there must be a filter(s) placed on the "Sun side" of the optics. For example, using welder's goggles over your eyes while looking through an unfiltered pair of binoculars can be dangerous as the goggle's glass could heat up and crack. Also, the attenuation may be inadequate for the focused, concentrated light coming through the lens.

You can purchase pre-made filters to fit your binoculars or camera lens on the Internet. Simply measure the outer diameter of the tube holding your lens and purchase a filter with a slightly larger diameter. The filter will slide over your exiting tube and will be held in place with either a foam or felt lining inside the filter tube. See Chapter 12, "Resources" for sources of pre-made filters and solar filter sheets.

Examples of pre-made solar filters suitable for binoculars and cameras, available in various sizes. Courtesy of Seymour Solar (glass) and Thousand Oaks Optical (black polymer).

It is very simple to make home-made filters that fit your binoculars (or camera lens) from specially made solar safety film that is readily available from websites, including Amazon. Search for Baader AstroSolar Safety film (other manufacturers are also available) and then either fashion something out of cardboard, some Home Depot parts or use camera adapter rings. Camera UV filters can also be screwed together in a stack to hold the safety film. Instructions for making your own filters can be found on the Internet. See Chapter 12, "Resources" for websites.

Homemade filters using plastic pipe and Baader film. Left: Sun side facing up. Right: Foam inner ring for pressure fit to binocular barrel. Courtesy Rick Hull.

Homemade filters mounted on 7x50 binoculars. Courtesy Rick Hull.

It is the responsibility of the reader to ensure proper precautions are taken when viewing the eclipse. All reasonable attempts were made to assure correctness of information, however the Author and Publisher are not responsible for the accuracy, efficacy, supplier and other recommendations or omission of advice and related information with regard to eye safety herein.

Chapter 9. For Romantics:
Eclipses in Literature, Song, Science, and Antiquity

Eclipses have been interpreted as mystical events, portents of bad news, the subject of songs, a means to manipulate the ignorant, a convenient plot line, a metaphor for blind love, and a tool for scientific research.

Throughout human history, total solar eclipses have been considered especially mysterious. The dedicated storytellers, explorers, scientists, songwriters and priests who write about them have left their mark on our culture. These are their stories.

Contemporary Culture, Literature and Song

Shakespeare's Othello

An author can't go wrong beginning a chapter with the words, "romantics" and "literature" in the title, with a quote from William Shakespeare.

Othello

"Yes: 'tis Emilia. By and by. She's dead.
'Tis like she comes to speak of Cassio's death.
The noise was here. Ha! no more moving?
Still as the grave. Shall she come in? were't good?
I think she stirs again:—no. What's best to do?
If she come in, she'll sure speak to my wife:
My wife! my wife! what wife? I have no wife.
O insupportable! O heavy hour!
Methinks it should be now a huge eclipse
Of sun and moon, and that the affrighted globe
Should yawn at alteration."[27]

Othello overhears a conversation, misinterprets what someone says, and thinks his wife (Desdemona) has cheated on him. He smothers her with a pillow, only to be caught by the handmaiden Emilia. He is arrested and upon learning of his tragic mistake, in a fit of despair over the loss of his love, stabs himself to death.

In ancient times, it was believed that a tragic human event (like the death of a king) would trigger celestial events. Othello is surprised that there is no eclipse since his wife had been so important in his life.

"Total Eclipse of the Heart" by Bonnie Tyler

"Once upon a time
I was falling in love
But now I'm only falling apart
There's nothing I can do
A total eclipse of the heart

Once upon a time there was light in my life
But now there's only love in the dark
Nothing I can say
A total eclipse of the heart."

This song was written and produced by Jim Steinman and recorded by Bonnie Tyler on her fifth studio album, *Faster than the Speed of Night* (1983). "Total Eclipse of the Heart" became Tyler's biggest career hit, reaching number one in several countries including the United Kingdom and the United States. This made her the first and only Welsh singer to reach the top spot of the Billboard Hot 100.

It was Billboard's number six song of the year for 1983. With physical sales in excess of nine million copies, Tyler's recording is one of the best-selling singles of all time. The song has been certified Platinum for U.S. shipments of more than two million copies by the Recording Industry Association of America (RIAA).[28]

"You're So Vain" by Carly Simon

"Well, I hear you went to Saratoga
And your horse, naturally, won
Then you flew your Learjet up to Nova Scotia
To see the total eclipse of the Sun
Well, you're where you should be all the time
And when you're not, you're with some underworld spy
Or the wife of a close friend,
Wife of a close friend, and

You're so vain
You probably think this song is about you..."

Written and performed by Simon on her third album, *No Secrets*, (released in November 1972). It is ranked #82 on Billboard's Greatest Songs of All-Time, and is considered her biggest hit and signature number. The song is memorable for its clever self-reference and the ongoing speculation about who Simon had in mind when she wrote the lyrics.

The song includes the following reference: "Then you flew your Lear jet up to Nova Scotia to see the total eclipse of the Sun." The event to which she refers is either the March 7, 1970 Eclipse or the one on July 10, 1972. Both were visible from Nova Scotia. Although the July '72 eclipse occurred eight months after the album's release, Simon may have used the past tense for consistency with the rest of the lyrics.

For 47 years, Ms. Simon has been dropping hints and telling celebrities under personal oath, the name of the person who is the subject of the song. In her memoir "Boys in the Tree" released in 2015, she finally revealed that the second verse was about Warren Beatty.[29]

Henry Wordsworth "The Eclipse of the Sun," 1820

HIGH on her speculative tower
Stood Science waiting for the hour
When Sol was destined to endure
'That' darkening of his radiant face
Which Superstition strove to chase,
Erewhile, with rites impure.

Sees long-drawn files, concentric rings
Each narrowing above each;—the wings,
The uplifted palms, the silent marble lips
The starry zone of sovereign height
All steeped in this portentous light!
All suffering dim eclipse!

Lo! while I speak, the labouring Sun
His glad deliverance has begun:
The cypress waves her sombre plume
More cheerily; and town and tower,
The vineyard and the olive-bower,
Their lustre re-assume!

Three selected stanzas from "The Eclipse of the Sun."

The Lunar Eclipse that Saved Columbus

Christopher Columbus could not get sponsorship in his native Italy to seek out a shorter, direct route to the Far East. The Spanish crown was better organized than its Italian counterpart and King Ferdinand and Queen Isabella agreed to fund the Explorer. Columbus set sail from Seville Spain in 1492. This was the same year these monarchs launched the Spanish Inquisition, killing tens of thousands and forcing out Muslims, Jews and other non-Catholics from Spain.

Columbus landed in the Caribbean islands, discovering the New World. He and his men became stranded on the island of Jamaica. Starvation threatened, but natives supplied them with food to survive until the sailors began stealing from the locals. After six months on the island, the natives suspended the life-sustaining food supplies.

Columbus had an almanac authored by Abraham Zacuto, which contained astronomical tables covering the years 1475–1506. Upon consulting the book, he noticed the date and the time of an upcoming lunar eclipse, February 29, 1504. Columbus told the native leader that god was angry and would provide a sign of his displeasure by making the rising Full Moon appear "inflamed with wrath."

The lunar eclipse and a red Moon appeared on schedule. Ferdinand, the son of Columbus recorded the event:

"[The People] with great howling and lamentation they came running from every direction to the ships, laden with provisions, praying the Admiral to intercede by all means with God on their behalf; that he might not visit his wrath upon them."

Columbus went into his cabin to "pray," emerging shortly before the eclipse ended. He told the frightened indigenous people that they were going to be forgiven and that God had pardoned them. Supplies resumed, and Columbus returned to Spain, making history.

Logs confirm that Columbus had extensive knowledge of celestial navigation and used a lunar eclipse to measure his longitude during the voyage. Those measurements were in error and served to help convince him that he had reached China although he had actually discovered an entirely unknown continent. Interestingly, throughout their remaining lives, Columbus, Ferdinand and Isabella all thought they had discovered a new route to the Far East. They never realized Columbus had discovered a new world—later to be named America.

Over the years, this Jamaica incident inspired subplots in several novels, including Mark Twain's "A Connecticut Yankee in King Arthur's Court" (a solar eclipse). A similar plot was also deployed in H. Rider Haggard's 1885 novel, "King Solomon's Mines" (a lunar eclipse).

A Connecticut Yankee in King Arthur's Court

In Mark Twain's 1889 novel, "A Connecticut Yankee in King Arthur's Court," the character Hank Morgan is struck on the head, traveling back in time to King Arthur's Camelot in the year 528 AD. Hank uses his knowledge of modern technology (from Twain's time of the 1880s) to gain influence and improve medieval living conditions. Merlin, the court magician, joins the townsfolk in ridiculing Hank for his strange appearance and manner. Hank is sentenced to burn at the stake, but remembering a solar eclipse is about to occur, he tells the king he will blot out the Sun if they continue to try and execute him. The eclipse occurs, Hank is pardoned, and he becomes principal minister to the King.

The main character takes advantage of the ignorance of the locals and convinces them he controls great power by removing, and later restoring the Sun. Twain's fictional plot mechanism was inspired by the true history of Christopher Columbus' experience while shipwrecked in Jamaica.

Although not the first story to make use of time travel, Twain's book came six years before the publication of H.G. Well's 1985 seminal novel, "Time Machine." Connecticut Yankee inspired several derivative works, including a Bing Crosby movie, a Broadway musical and the Warner Brother 1979 cartoon special, "A Connecticut Rabbit in King Arthur's Court."

Eclipse, the Racehorse

The racehorse "Eclipse" was considered to be the finest racehorse of the late 18th century. He was so named because of his birth on April 1, 1764 during an annular solar eclipse. Using DNA testing, scientists have recently confirmed that 95% of all thoroughbred racehorses are directly descended from this one animal. In the U.S., the "Eclipse Award," named after this horse, is bestowed upon thoroughbred champions.[30]

Eclipses in Modern Science

The Harvard Expedition

During the Revolutionary War, Harvard University organized the first American eclipse expedition with the goal to observe the October 27, 1780 Total Solar Eclipse. Unfortunately, the only site with sufficiently deep water to allow a ship large enough to carry the necessary equipment lay in British-held territory. John Hancock, the speaker of the House, pleaded with the British commander at Penobscot Bay to make the American expedition possible. Addressing the commander as a "Friend of Science," Hancock wrote: "Though we are political enemies, yet with regard to Science it is presumable we shall not dissent from the practice of all civilized people in promoting it . . ." A special immunity agreement was negotiated with the British so that the scientists could work unharmed. The Harvard expedition, after all their efforts, didn't see the eclipse because they accidentally chose a site outside the path of totality.[31]

Discovering Helium

About 98% of the entire universe is made up of two elements, hydrogen (74%) and helium (24%). All the other elements (oxygen, carbon, iron, gold, etc.) make up the remaining 2%.

A nuclear fusion source like the Sun emits a broad spectrum of electromagnetic radiation, including light. In 1859 it was theorized that a gas (or plasma) between the Sun and the Earth would absorb specific wavelengths of light and this would show itself in the spectrum arriving here on Earth.

A refresher: What is a *spectrograph*? White light is created by adding together all the colors of our visible spectrum, from red to violet and all the colors in between. We can separate the light into its constituent colors using a prism or a *diffraction grating*. Certain elements (such as hydrogen or helium) will absorb *photons* of light and change the spectrum by removing (or sometimes reinforcing) certain colors (*wavelengths*). This will show up as either black lines in the split apart spectrum or by added brightness to certain colors. Scientists use this technique of *spectroscopy* to identify what elements are in a material.

During the eclipse of August 18, 1868, the French astronomer Pierre Janssen observed totality from Guntur in Madras State, British India. It was the first total eclipse since the 1859 theory that the lines in the solar spectrum correspond to the emission line of the different chemical elements present in the Sun. Observing with a spectrograph, Janssen noticed a bright yellow line in the spectra of the solar prominences. The same result was found independently by British astronomer Norman Lockyer, and both Janssen's and Lockyer's communications were presented to the French Academy of Sciences on October 26, 1868. Lockyer named the new element Helium, after the Sun god Helios. It is interesting to note that Helium was found to exist in the atmosphere of the Sun before it was identified here on Earth.

The present day largest single commercial use of helium (in liquid form) is to cool the superconducting magnets inside MRI scanners. It is also used for various research projects at very low (*cryogenic*) temperatures, arc welding and in commercial processes such as growing crystals to make silicon wafers.

Other well-known uses are as a lifting gas in balloons, and making us sound like Donald Duck at birthday parties.

Confirming the Theory of Relativity

According to Einstein's 1915 "General Theory of Relativity" (his "Special Theory of Relativity" was published earlier in 1905), gravitational fields should cause space to bend (warps in space-time). Light should therefore not continue in a straight line, but follow the curve of space near a large gravitational field. Einstein was a theoretician steeped in the world of mathematics and proofs, so experimental physicists were left to verify his theory. To bend space on a large enough scale to measure its curve, requires a huge gravitational field. Such a field is generated by the mass of the Sun.

At night, the light on the way to Earth from far off stars does not pass close to the Sun. During daylight hours, a star's light might pass close to the Sun but we can't see it due to the Sun's brightness.

To conduct this experiment, scientists chose a star that appears close to the Sun during a total solar eclipse. During totality they measured its distance from a reference point (another star). They next measured the same star's position during the night when its light did not pass near the Sun. If there was a shift in the star's apparent relative position, it must have been caused by the warp of space-time predicted by Einstein.

Mathematically the theory predicts that a ray of light will shift very slightly, less than two *arcseconds* as it passes near the Sun. An *arcsecond* is 1/3,600 of a degree which is the angle subtended by a US dime located at a distance of about 2.5 miles away. This is a very small angle. Imagine a right triangle that is one inch at its base and about two miles tall. The angle at the top is about two arcseconds.

A team of scientists lead by Sir Arthur Eddington and his collaborator Sir Frank Watson Dyson traveled to the island of Principe off Africa to observe the May 29, 1919 total solar eclipse.[32] The Sun would be passing close to the relatively bright Hyades star cluster during the eclipse. Totality was particularly long for this eclipse at 6 minutes and 51 seconds.

The results matched theoretical predictions and these eclipse measurements became the first widely accepted proof that Einstein was correct. The news was reported in all the major newspapers, making Einstein and his theory world famous. Over the years, this measurement was repeated by others, including several world renowned astronomy teams during the 1922, 1953 and 1973 eclipses.

May 29, 1919 Eclipse. Photo from the report of Sir Arthur Eddington on the Expedition to the island of Principe (off the west coast of Africa). Public domain.

Eclipses in Antiquity

Ancient records exist referencing eclipses in Babylonia and China as far back as 4,000 years ago. Eclipses of the Sun and Moon have always left a profound impression on witnesses. When the ever-present Sun was suddenly taken from the sky, or the Moon turned red as if covered in blood, there was a need for an official explanation and some action. Losing the Sun was considered a bad omen.

Thousands of years ago, the ability of a kingdom to predict both lunar and solar eclipses was often regarded as essential to maintaining political order. Investments were made by the great ruling courts of old in astronomical understanding and prediction. Citizens needed to be calmed, and Kings needed to be reassured these events did not mean the gods were angry at their rule.

Using computers, we have verified that ancient eclipses did, in fact, happen at the times stated in these records. While Fred Espenak was at NASA, he published a compendium of all solar and lunar eclipses going back to 2,000 BCE ("Five Millennium Catalog of Solar Eclipses (-1999 to +3000)" and the "Index to Five Millennium Catalog of Lunar Eclipses (-1999 to +3000").

The Chinese

The ancient Chinese believed the eclipse was caused by a dragon who was attempting to devour the Sun. To frighten away the dragon, locals would make noise, beat drums and launch arrows at the sky. It was thought that an eclipse foretold the future of the Emperor, so being able to predict the event beforehand was of some importance.

One of the earliest written records of a total solar eclipse comes from China. The classic tale "Shu Ching" tells the story of two court astronomers who had their heads removed for failing to foretell the solar eclipse of October 22, 2134 BCE. Predicting an eclipse was a duty of ancient Chinese astronomers. The two royal astronomers, Hi, and Ho, knew that an eclipse was imminent. According to legend, they were drunk from an anticipatory celebration of the rewards and gifts the emperor would bestow upon them. Time slipped away and on the day of the eclipse they couldn't perform the rites of chanting, beating drums and shooting arrows to chase away the dragon. When the eclipse took place, the emperor—also known as the "Son of the Sky"—was caught unprepared. The emperor ordered Hi and Ho beheaded for their transgressions.

By 20 BCE, Chinese astronomers understood the true nature of an eclipse and around 200 CE, they were able to reliably predict future events by modeling the motions of the Moon.

Stonehenge

Built around 1900 BCE, in Southern England stands an awesome arrangement of prehistoric ruins and stones that have been the subject of countless studies, poems and legends. Speculation on the study of Stonehenge have continued unabated from the time that it was first mentioned in the literature shortly after the Norman Conquest (1066). Evidence indicates that Stonehenge, built during the same era as the Great Pyramid of Egypt, was a brilliantly conceived astronomical observatory. Certain features were apparently used as an eclipse predictor.

Egypt, Mesopotamia and Babylon

From Mesopotamia and Babylon, we have records of the eclipse of May 3, 1374 BCE in the city of Ugarit, a port city in Syria. Also recorded was the total solar eclipse on July 31, 1036 BCE. The Babylonian astronomers observed the motions of the Sun, Moon and planets. They kept careful records of celestial events. Egyptian tombs indicate that the solar and lunar eclipses were observations in the ancient kingdom.

Bible and Assyrian Tablets

In the Hebrew Bible, reference is made to a total solar eclipse that is believed to have occurred on June 15, 763 BCE in the city of Nineva (present day Iraq). An Assyrian tablet also verifies the event. By aligning the Bible story with Assyrian writings, historians have been able to pinpoint dates for several key biblical events.

Greek Astronomers

Around 230 BCE, Aristarchus made an estimate of the diameter of the Moon and proposed the first model of the solar system with the Sun at its center. About the same time, Eratosthenes used the angle of shadows cast at noon in Aswan and Alexandria on the summer solstice to estimate the circumference of the Earth. In 130 BCE, Hipparchus determined that the Moon was 268,000 miles away, within 11% of today's accepted distance.

"Ptolemy (150 BCE) represents the epitome of Grecian astronomy, and surviving records show that he had a sophisticated scheme for predicting both lunar and solar eclipses. Ptolemy knew, for example, the details of the orbit of the Moon, including its nodal points—and that up to two solar eclipses could occur within seven months in the same part of the world." From NASA eclipse article, see endnote for reference.[33]

The Greek Antikythera Mechanism

In 1901 a complex ancient analog mechanical computer dating back to 100-150 BCE was discovered near a shipwreck off the Greek island of Antikythera. The mechanism has over 30 meshing bronze gears and was used by ancient Greek astronomers to predict celestial events including lunar and solar eclipses.

The Crucifixion of Jesus

The Christian gospels state that the sky darkened after Jesus' crucifixion. Two possible eclipses around that time may have been responsible, either the eclipse of 29 CE or 33 CE.

Birth of Mohammed

The Koran mentions an eclipse occurring just prior to the birth of Mohammed. This may have been the total solar eclipse 569 CE.

Total Solar Eclipse 2017

Chapter 10. The Motions of the Sun, Earth and Moon:

Answers to Advanced Questions

Together, Chapter 4 ("Up There: How Eclipses Happen") and Chapter 5 ("Down Here: How the Shadow Moves Across the Earth") provided a solid foundation to understand how eclipses occur. This chapter, "The Motions of the Sun, Earth and Moon" is extra credit. It fills in the blanks, completing your understanding of eclipse science. This chapter anticipates and answers the following questions:

What are the Moon's phases and why do they occur?

How do the planes of the Earth and Moon orbits interact to produce an eclipse?

Do eclipses repeat and how did the ancients predict them without computers?

What exactly determines the direction and speed of the eclipse shadow?

The Moon's Phases

The Moon does not radiate light; it only reflects light coming from the Sun. The "phases of the moon" are simply the various ways the Moon appears to us throughout the month due to the angle at which sunlight strikes the lunar surface.

1. When the Moon is between the Earth and the Sun, only the back side of the Moon is illuminated. But we can't see the back side because it is facing away from us. The side of the Moon facing us appears dark, making the Moon seem to disappear. This is the *New Moon*. To the naked eye, it is invisible but a telescope collects enough stray light bouncing back off the Earth that we can still see the Moon as a dimly lit disk in the sky.

2. Looking down from space, high above Earth's North Pole, the Moon travels counterclockwise around the Earth. This is the same direction the Earth spins on its axis. When the Moon gets one fourth of the way around (90 degrees), we see it half-lit per the figure below. This is the *First Quarter Moon*.

3. When the Moon reaches the position opposite the Earth and the Sun (180 degrees), we see the completely lit lunar disk, this is the *Full Moon*.

4. At the three-quarters mark (90 degrees), the Moon's disk again appears half-lit. This is the *Third Quarter Moon*.

5. The Moon then returns to its *New Moon* position.

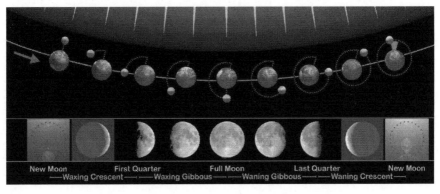

Location of the Moon and its appearance seen on Earth.
Source: "Lunar phase," Wikipedia, by Orion 8.

Lunar phase terminology can be confusing; here is how it works. The word *crescent* refers to the phases where the moon is less than half illuminated. The word *gibbous* refers to phases where the moon is more than half illuminated. *Waxing* means "growing" or expanding in illumination, and *waning* means "shrinking" or decreasing in illumination.

To name the phase of the Moon, you combine the two words. For example, after the New Moon, the sunlit portion is increasing, but it is less than half, so it is a *waxing crescent Moon*.

A solar eclipse only occurs when there is a New Moon. But a New Moon is invisible because the side of the Moon facing us receives no light. From this discussion, it follows that while observing a solar eclipse, you can't see the Moon. This is correct. We don't see the gray disk of the Moon moving over the bright Sun. Instead, it appears to us as if something invisible is simply taking bites out of the Sun—the Pac-man effect mentioned earlier.

How long does it take the Moon to go around the Earth? There are two periods involved with lunar orbits, and this can lead to some confusion. For this discussion, keep in mind our calendar month is approximately 29.5 days long (a *synodic period*).

The time for the Moon to make a complete orbit of the Earth is 27.3 days (the *sidereal period*). You may be surprised that this is shorter than a month. Our month is based on how long it takes for the Moon to return to the same position in the sky relative to the Sun. In this manner, a New Moon will occur once per month, as will all the subsequent lunar phases.

The Moon takes 29.5 days to return to the same point as referenced to the Sun because over the course of the lunar orbit period (27.3 days), the Earth has moved forward in its orbit around the Sun. The Moon must catch up with this movement, and this takes an extra 2.2 days.

Earth and Moon Planes of Orbit

In the beginning, the planets formed out of a disk of dust that surrounded the Sun. This led to the planets, including the Earth, all being created in a more-or-less flat, fixed plane. The orbits of all the planets in our solar system lie very close to this same plane, called the *ecliptic*. Our Moon however, orbits the Earth in a plane that is tilted five degrees in reference to the *ecliptic plane*.

When two orbits are tilted relative to each other, the orbital planes will cross exactly twice at points called *nodes*. The Moon orbits Earth once every month; therefore, the Moon's orbital path crosses the ecliptic plane twice each month at the *ascending node* and the *descending node*. The figure below helps visualize the two nodes:

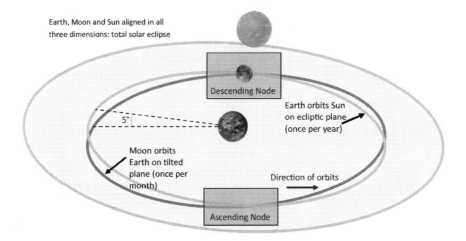

Total solar eclipse: The Moon crosses the plane of the ecliptic twice each month. Solar eclipses occur when this crossing coincides with the Moon being located on a line between the Earth and the Sun (a New Moon). We see a total solar eclipse if the Moon is close enough to Earth, otherwise the eclipse will be annular.

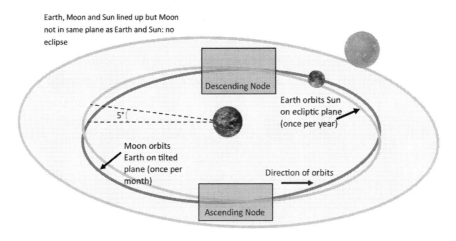

No solar eclipse: The Moon, Sun and Earth are lined up in the "New Moon" phase but the Moon is well outside the ecliptic plane.

For a solar eclipse to occur, the Moon must lie in a straight line between the Sun and the Earth (horizontally lined up as during the

New Moon phase) while at the same time, the Moon is crossing the ecliptic plane (vertically lined up). Only then are the Earth, Moon and Sun aligned in three dimensional space. When this alignment occurs we get a solar eclipse.

A *node* aligned solar eclipse (also called a *central solar eclipse)* occurs when the New Moon occurs precisely coinciding with a node crossing. *Central eclipses* can be either annular, (when the Moon is far away and appears smaller) or total, (when the Moon is close and it appears large enough in the sky to block the Sun).

If the New Moon does not coincide precisely with a node crossing, but falls within 18.5 degrees of one, a partial solar eclipse will be seen.

On average, a total solar eclipse occurs about once every 16 months (includes hybrid eclipses). In addition, an annular eclipse occurs once every 15 months and a partial solar eclipse happens once every 14.3 months.[34]

Eclipse seasons: The Sun will come within 18 degrees of one of the Moon's crossing nodes twice every year, about six months apart. Astronomers refer to the approximately 37 days around each of the crossings as "eclipse seasons." Eclipse season in 2015 occurred in late March–April and then six months later in September. All eclipses in any given year (lunar and solar) will occur during this twice yearly "eclipse season."

Eclipse Cycles: The Saros

Since celestial motions are repeatable events, it seems intuitive that eclipses should repeat themselves over time; in fact, they do. All the individual conditions that set up one eclipse will, at some point, come around and cause another similar eclipse. If you simulate on a computer past and future eclipses based on celestial mechanical models, you will indeed see repeating patterns.

It did not take computers for astronomers to notice these patterns. About 2,800 years ago, Babylonian astronomer-priests identified what is known as the "Saros cycle." Saros is derived from the Greek root saru (repetition). The Saros cycle is equal to 18 years, 11 1/3 days. In every Saros cycle, the relative positions of the Sun and Moon and the orbital nodes closely repeat themselves. Because of the one-third of a day in the cycle, the earth has rotated by one-third day between eclipses within a Saros cycle. Therefore, the repeating eclipse does not fall on the same portion of the globe each time, but moves around the Earth in a westward direction. Over three cycles (approximately every 54 years) the eclipses return to approximately the same location.

A set of related eclipses (spread over centuries) makes up one Saros series. There are about 40 Saros series active at any point in time. Each series starts over one of the poles and then shifts slightly towards the opposite pole with each event. A Saros series can last between 12 and 13 centuries and include as many as 87 solar eclipses of all types[35]. The total solar eclipse on August 21, 2017 is a member of Saros 145 that began on January 4, 1639. The last time this Saros produced an eclipse was August 11, 1999, a total eclipse centered over the UK and the European continent. The next Saros 145 eclipse (after 2017) will deliver totality over China and Japan on September 2, 2035. Other repeating eclipse patterns have been identified, but the Saros cycle is the most useful and most often cited.

Direction and Speed of the Eclipse Shadow

Things are always moving during an eclipse. The shadow is moving through space because the Moon is orbiting the Earth. The Earth is also revolving on its axis. These motions combine to create the motion of the shadow here on Earth.

Because the Earth rotates on its axis and because the Earth is orbiting the Sun, as we stand on the Earth's surface you and I are hurling through space at thousands of miles per hour. We don't feel this motion because everything around us (buildings, trees and the ground) are all moving at the same speed.

Since we are moving relative to the Sun and Moon, the following question comes to mind: At what speed will the shadow appear pass over the Earth and in what direction?

First consider the following: If you are driving your car at 60 miles per hour and a motorcycle moving in the same direction passes you at 80 miles per hour how fast does the motorcycle appear to be moving past you? Answer: 20 miles per hour.

The Earth rotates on its axis once every 24 hours. This means a person standing at the equator will have a linear velocity through space of 1,044 mph (radius of Earth x 2 x pi / 24 hours) = 1044 mph). This speed decreases by the cosine of your latitude so that at a latitude of 33.6839 degrees (Irvine California) our speed due to the rotating Earth is 0.642 x 1044 = 670 mph. The cosine of 90 degrees = 0, therefore a person standing at one of the Earth's poles has zero linear motion contribution from the Earth's rotation. This makes intuitive sense if you recall the Roundabout playground you played on as a child. The further towards the outer edge the faster you went. Move towards the center and you move slower in a tighter circle.

Now consider the motion of the Moon. The Moon's orbit is gravitationally linked to the Earth's rotation. Therefore, it should come as no surprise that the Moon orbits the Earth in the same direction as the Earth's spin; counterclockwise (as viewed from space, above the North Pole). The Moon orbits once every 27.3 days, which works out to 2,274 miles per hour, the same as the Moon's shadow.

Let's put this all together. The Moon's shadow is moving at 2,274 miles per hour on an earth that is spinning in the same direction at 1,044 miles per hour at the equator. Therefore an observer standing at the equator will see the Moon's shadow move across the Earth at the rate of 1,230 miles per hour (2,274-1,044). If our observer is standing at the North Pole, our observer will see the Moon's shadow moving at

2,274mph (2,274-0). This means the speed of the shadow will vary based on latitude from 1,230 miles per hour along the equator to over 2,000 miles per hour near the Poles.

Now consider the direction of the shadow as it sweeps across the globe. For this analysis you can simplify things by imagining that the Earth is stationary and the only motion is the Moon orbiting Earth. This simplification makes sense since the eclipse takes place in just a few hours, not over multiple days. From the picture below, it is obvious that the shadow will sweep eastward (for example from Los Angeles to New York) across the globe. The shadow's motion will also include a considerable north-south component depending upon the exact Moon-Earth-Sun geometry.

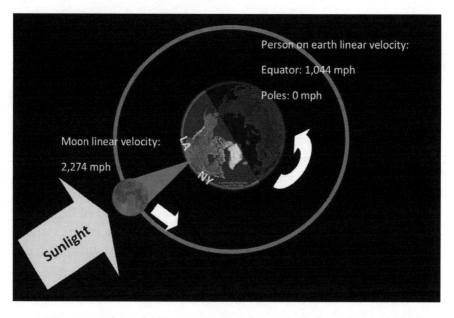

Direction and speed of rotation as seen from space, above the north pole:
The shadow sweeps eastward at speeds of 1,230 mph at the equator and
up to 2,274 mph at the poles.

What does all this mean to us? The eclipse shadow will always move across the globe from west to east and in 2017, if you are standing on a hill in the United States, you will see the Moon's shadow rushing towards you at over 1,500 miles per hour!

Chapter 11. Photographing the Eclipse:
Tips for Smartphone Cameras, SLRs and More

Scenes to Be Captured

No two total eclipses are the same. How high the Sun is in the sky, the sunspot cycle, the weather, and your observation location are just a few of the factors that play a role in making each one unique. Sometimes the corona is tight and circular, sometimes it's wispy and petal-like. The point is, you will want to see "your" eclipse in pictures afterward. Looking in a science book at an image of an example total solar eclipse is just not the same as seeing an image of the eclipse you trekked out to observe.

Photographing a total solar eclipse involves capturing three vastly different scenes:

> 1) Photographing the people. You may want to capture the reaction of friends and family during set-up, the partial phases, as totality approaches and just after totality ends. Don't forget to grab a quick shot of mesmerized friends staring upward during totality while their protective glasses are off. Seconds after totality ends I can pretty much guarantee some of the biggest, natural smiles of your photographic career! This is my favorite part of the event: taking pictures of people. It's also a great time to video a spur of the moment reaction to totality. Ask them "what did you think" and let the camera roll.

> 2) Photographing the Sun before and after totality. Requires filters for your camera.

> 3) Photographing the Sun during totality. No filters required.

If this trip will be your first time viewing totality, I highly suggest you either just try to capture images of people, or if you are a bit more ambitious, take some pictures during during the partial phases of the eclipse. During totality it is best to not be fiddling with your camera. You may be able to get copies of totality pictures from the people around you or on the Internet later.

Most film cameras and some digital cameras have optical viewfinders. The same way you don't ever want to look directly at the Sun outside of totality, you can also do serious and permanent damage to your eyes if you look at the sun through an optical viewfinder. This damage can happen very quickly because the lens acts to concentrates the light. A filter designed for direct solar viewing is necessary to protect your eyes and is required for a correct exposure during the partial phases.

Many digital cameras can present an image of what the lens is seeing on the LCD screen on the back of the camera. This is sometimes called "live mode" or "live view." While you may be able to safely watch your camera's electronic display, without a proper lens filter you can damage the camera's sensor.

General Camera Notes

Because the camera market has gone almost entirely digital, I will not address film cameras. Digital cameras capture the image with silicon sensors and store the resulting image file on internal solid-state memory or memory cards. Although some of these cameras can be attached to telescopes, our interest here is their use as stand-alone instruments. Selecting and using astronomical telescopes to capture images are beyond the scope of this book.

Any time you point a camera directly at the Sun during the partial phases of the eclipse you will need to protect your camera with a proper solar filter. Without a filter, your camera's sensor can be quickly destroyed, and the heat from long exposures can warp precision parts. Also, if your camera has an *optical* viewfinder (in addition or in place of an electronic display*)*, you will damage your eyes if a proper filter is not used. With an optical viewfinder, never look at a Sun that is not totally eclipsed using just your solar safety glasses, there must be a proper solar filter on the sun-side of the camera lens. Do not use standard or even very dark ND (neutral density) filters as these were not intended for direct observation of the Sun.

Before we dig in further, I offer the following advice: whenever photographing the Sun during an eclipse, bring along a camera mount. I suggest either a tripod for removable lens cameras or something smaller for point and shoots and smartphones. The mount will make things easier and it's essential if you want to create sharp images.

Photographers attempting to capture the partial and totality phases of any eclipse face three challenges: 1) The size of the Sun in the frame, 2) getting the exposure correct (light or dark) and 3) achieving a sharp, in-focus image. We will discuss each of these in the next few sections.

Classes of Cameras

There are many ways of slicing and dicing the camera market. Lens size (also called focal length or zoom level) and the size of the *sensor* are the most significant factors that determine image quality and usefulness for taking photos of the eclipse. The *sensor* is a silicon chip that acts like a piece of film. The lens focuses the image onto the *sensor*, and the *sensor* converts the light into electrical bits and bytes that are stored in the camera's memory card. In general, the larger the *sensor*, the better the image quality. Also, on a particular camera, the Sun will appear larger in the frame using a longer focal length lens, resulting in a better rendering of fine details.

For the purpose of this book, I assume you have one of three types of cameras:

1. Non-interchangeable lens camera: This category includes DSLR-like (Digital Single Lens Reflex) compact cameras with powerful zoom lenses and simple point and shoots with modest zoom lens capability. In addition to having a lens that is permanently attached, all these cameras use a smaller class of image sensor than higher-end cameras. There is no standard sensor size in this category. Therefore, manufacturers state the capability of their cameras as being "equivalent to" a full frame camera with an interchangeable lens (see #2 below). In effect, you don't need any knowledge about the sensor size or a number called *crop factor*. All you need to know is the manufacturer's stated focal length. A typical point and shoot camera specification might read "4x lens from 25mm wide angle to 100mm zoom *equivalent*."

2. Interchangeable lens camera: These units use one of three size sensors named Full Frame (FF), Advanced Photo System type C (APS-C) and Four-Thirds (FT). It should be noted that there are slight size differences between the APS-C sensor used by Nikon/Sony and Canon. The Full Frame camera is the gold standard and has been assigned a 'crop factor' of 1. Cameras with APS-C sensors have a crop factor of either 1.5 (Nikon/Sony) or 1.6 (Canon). A camera with a Four-Thirds sensor has a crop factor of 2.0. For a complete discussion of crop factor and how it relates to the size of the Sun's image in your interchangeable lens camera's frame, see the section "Image Size for Interchangeable Cameras" later in this chapter.

3. Smartphone camera: These use the tiniest sensor chips, and all have fixed size lenses.

Challenge 1: Image Size

Getting an image of the Sun large enough for a good picture is our first challenge. For a high quality image with detail, target a minimum of about 25% of the camera frame filled by the Sun. If you go too far over 50%, visible portions of the corona might be cut off. Photographing close ups of the solar prominences is a special case, and you may want considerably more magnification.

Lens Field of View (FOV) and Focal Length: From Earth, the Sun and the Moon both appear the same size in the sky, about one-half of a degree. What does this mean? In this measurement system, 360 degrees is a full circle, and 180 degrees is a half circle. Humans have a horizontal *field of view (FOV)* of 180 degrees. We can't see past this without turning our heads. The eyes of some birds, however can see almost 360 degrees without turning their heads.

When we say the Sun has a "size" of one-half degree, we mean that out of all the 180 degrees humans can see when looking forward, the Sun appears to take up only half of one of the 180 degrees. That's pretty small.

Let's relate this back to cameras and lenses. If we had a camera with a field of view of 180 degrees (an extreme wide angle or fish eye lens), a picture taken during an eclipse would render the Sun as a tiny dot, taking up 0.5/180 or about 0.3% of the frame.

Now we'll take this same picture with a telephoto lens that has a *field of view* of five degrees. The Sun would be a circle of 0.5/5 or 10% of the frame.

If we want the Sun to fill one-quarter (25%) of the height of the frame, we need a camera-lens combination with a vertical FOV of about two degrees (0.5 degrees/2 degrees = 25%). On a typical non-interchangeable lens camera (or a full frame camera), we would need a 700mm focal-length telephoto lens. Lenses with focal lengths above about 85mm are commonly called telephoto lenses.

Image Size for Non-Interchangeable Lens Cameras (Advanced and Standard Point-and-Shoot)

Point-and-shoot cameras come in a variety of sizes targeted to various user skill levels. Some fit in a shirt pocket and others are almost the size of interchangeable lens DSLRs cameras. Almost all share the characteristic of small sensors, making them less than ideal for photographing the Sun. However, a few of the larger cameras in the class with powerful zoom lenses will deliver perfectly acceptable images of the eclipse.

To figure out how large the Sun's image will appear in the frame, you need to know the camera's "equivalent" (sometimes called "full frame equivalent") maximum focal length. The Canon PowerShot S95 has a zoom lens with an equivalent focal length of 28mm to 105mm. If you took the camera apart, you'd discover the actual measurement of the lens to be 6mm to 22.5mm in focal length. We don't care about this actual measurement; only the "equivalent" focal length matters. For this class of camera, printed specifications in the manual and on the lens are stated as equivalent to full frame focal lengths. You can look up 105 mm in the table below and find that the best the Canon PowerShot S95 will deliver is an image of the Sun that is about 4% of the vertical frame.

Ideally, I recommend an image of the Sun that fills between 25% and 55% of the vertical portion of the camera frame, as highlighted in the following table.

Lens Focal Length (mm)	Sun as % of vertical frame
100	4%
150	6%
200	7%
300	11%
400	15%
500	18%
600	22%
700	26%
800	29%
900	33%
1000	37%
1100	40%
1200	44%
1300	48%
1400	51%
1500	55%
2000	73%

Table of size of the Sun's image on typical non-interchangeable lens cameras.

Recommended range for full image of Sun and corona is highlighted.

There are a small handful of point-and-shoots with very high power telephoto lenses. Using the 25% to 55% guideline, a few of these cameras will deliver an acceptable size image of the Sun:

Canon PowerShot SX60 HS: 21mm–1365mm

Nikon Coolpix P900: 24mm–2,000mm

Sony HX300: 24mm–1,200mm

Panasonic FZ300K: 25mm–600mm (falls a bit short, with Sun filling 22% vertical)

This is the size of the Sun's image on a typical non-interchangeable lens camera display:

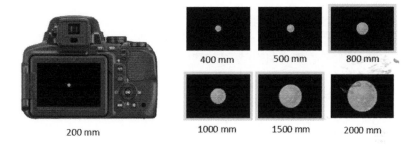

Sun size on typical non-interchangeable lens (advanced and standard point and shoot) cameras. Suggested focal lengths highlighted.
Nikon CoolPix P900 pictured.

If your camera can't deliver 25% vertical fill, don't give up. If you can fill at least 10% of the frame, by cropping and enlarging it on a computer, you should have a perfectly fine image for use on Facebook, vacation slideshows and other non-critical viewing applications.

Here is a cautionary tale: One time I used a small pocket point-and-shoot to snap a few pictures of an eclipse. The images looked acceptable on the back of the camera, however when I enlarged the image in Photoshop, instead of the Sun appearing as a more-or-less round ball, it had a rather bizarre oblong shape. I believe this was caused by distortion in the highly miniaturized zoom lens.

One downside to a smaller image sensor is that if you try to enlarge the Sun's image on a computer, there might not be enough resolution (pixels) to make an acceptable image.

In summary, if you intend to capture the Sun with a pocket-sized point-and-shoot, don't expect much. If your non-removable lens camera has a telephoto capability over 700mm (26%), you should get great images with fine detail. If focal length is between 300mm (11%) and 600mm (22%) with cropping and enlarging in Lightroom, Photoshop or your favorite photo editor, you will have acceptable images, depending on your final viewing environment.

Image Size for Interchangeable Lens Cameras

Digital cameras with interchangeable lenses use one of three different size image sensors, leading to three different levels of scaling between the focal length of a lens and the size of the image produced. Cameras such as the Nikon D4, D810, D610, D750, D4S, the Canon EOS 6D, 5D, 5DS, 1DX, the Sony Alpha a99 and Sony DSC-RX1 use a digital sensor that is the same size (35mm) as the film in our old film cameras. For this reason, cameras with a full frame sensor are considered the reference point and assigned a *crop-factor* of 1.0.

The majority of other digital SLR cameras have an intermediate sized sensor named "Advanced Photo System, type C" or "APS-C." These have a crop-factor of either 1.5 (Nikon, Sony, Pentax) or 1.6 (Canon). This means a 100mm lens on an APS-C sensor camera will deliver an image between 1.5 to 1.6 times larger than a 100mm lens on a full-frame camera.

Olympus and Panasonic Four Thirds and Micro Four Thirds mirrorless cameras have identical size sensors and a crop-factor of 2.0. Standard Four Thirds and Micro Four Thirds have different size lens mounts. A 100mm lens on these cameras will deliver an image two times larger than a 100mm lens on a full frame camera.

The following table takes sensor size and crop factor into account and lists the required lens focal length to achieve various image sizes of the Sun. Again, for high detail photographs I recommend an image size between about 25% and 55% of the vertical frame as highlighted in the table:

	Sun as % of Vertical Frame			
	Full Frame (FF)	APS-C (Nikon/Sony)	APS-C (Canon)	Four-Thirds
Crop factor -->	1.0	1.5	1.6	2.0
Lens Focal Length (mm):				
100	4%	6%	6%	7%
150	6%	8%	9%	10%
200	7%	11%	12%	14%
300	11%	17%	18%	20%
400	15%	22%	24%	27%
500	18%	28%	30%	34%
600	22%	34%	36%	41%
700	26%	39%	42%	47%
800	29%	45%	47%	54%
900	33%	50%	53%	61%
1000	37%	56%	59%	68%
1100	40%	62%	65%	74%
1200	44%	67%	71%	81%
1300	48%	73%	77%	88%
1400	51%	78%	83%	95%
1500	55%	84%	89%	101%
2000	73%	112%	119%	135%

Size of image of Sun with various lenses. Recommended range for full image of Sun and corona is highlighted.

This is the size of the Sun's image on typical interchangeable lens camera displays:

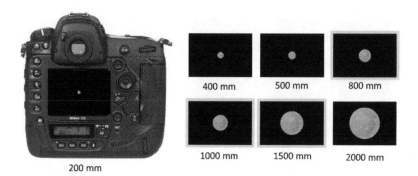

Sun size on full-frame cameras, suggested focal lengths highlighted. Nikon D4 pictured.

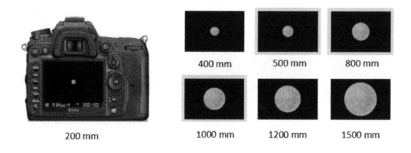

Sun size on APS-C cameras, suggested focal lengths highlighted. Nikon D7000 pictured.

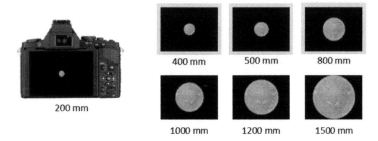

Sun size on four thirds cameras, suggested focal lengths highlighted. Olympus OMD EM-5 pictured.

For a given focal length lens, the field of view (FOV) will be different depending on the size of the sensor. The formula for determining the FOV of a lens/camera combination is: FOV = sensor size (mm)/focal length of the lens (mm) x (360/2 pi) or sensor size/focal length x 57.3. Since sensors are not square, cameras have different horizontal and vertical fields of view. For the table in this chapter, vertical FOV is used since we are trying to fill a percentage of the frame's vertical height. Since the field of view filled by the Sun and Moon is about 0.5 degrees, once you know field of view of a lens, the percent of the frame taken by the Sun's image is 0.5/lens FOV.

A high focal-length, super telephoto lens would be great to have on eclipse day, but you may not find much use for it in your everyday photography. That makes such a purchase a tough decision. Another approach is to use the lens you already own and crop your way to a larger image during post-processing on your computer. It all depends on how you will be viewing the final image. If you're aiming for a large print, a large focal length lens is necessary to bring out details.

However, if you intend to display the image on the web or in an HDTV slide show, the higher focal length lens may not be necessary. A camera with a high-resolution sensor (16 or 24 megapixels and up) should let you crop-to-enlarge in Photoshop or Lightroom with good results.

I created respectable eclipse images using a 150mm zoom lens on my Micro Four Thirds Olympus OMD EM-5 (Sun filled 10% of vertical frame). Using Lightroom, I cropped until the Sun filled about 50% of the frame. This image was good enough to use in Chapter 2 of this book (the diamond ring photo) and is good enough quality to post on Facebook. It also looked acceptable (but not fantastic) on an HDTV slide show shown on a 70-inch TV (from our watching position on the couch about 10 feet back from the screen).

The designation of *mirrorless* refers to a simplification of the time-honored SLR design. The SLR stands for Single Lens Reflex and the thing that was "flexing" was a mirror. In most SLR cameras, the mirror was used to send the image to the user's eye through the viewfinder to set up the picture. When the user pressed the shutter, a mirror flipped out of the way so the image from the lens would strike the film or digital sensor, completing the exposure. Mirrorless cameras send the image directly to the sensor, and the user "sees" what they are about to take a picture of through an electronic viewfinder or a small display panel on the back of the camera. Eliminating the mirror simplified things, reducing cost and weight.

Image Size for Smartphone Cameras:

The point-and-shoot camera market has rapidly declined as users have shifted to using smartphones for taking pictures. It's like they say, what's the best camera in the world? Answer: The one you have with you. The rapid expansion of smartphone sales has driven an explosion in picture taking. It is estimated that worldwide, about 100 billion photos were taken in 2000. For 2015, this is expected to reach 1.6 trillion photos.

The best smartphone cameras (on the newest Apple iPhones and Samsung Android phones) are now about the same quality as midrange point and shoots. However, phones don't have zoom lenses.

Most smartphones can zoom-in electronically (digital zoom). The number of pixels across your camera's sensor determines how well your camera can capture fine details. Digital zooming increases the pixels size, not the amount of pixels and the resulting image can lack clarity. It's best to perform digital zoom (cropping) on a computer to take advantage of the PC software's ability to fill-in missing pixels by interpolation (mathematical approximation). No matter which technique you use, too much digital zooming will result in a substantial loss of image quality.

Third party telephoto multiplier lenses are available that will double or quadruple the fixed focal length of your camera phone. The modest increase in focal length provided by these add-ons are not sufficient for taking close-up images of the eclipse.

Another approach is to use a pair of binoculars as a lens for your smartphone. Several manufacturers make adapters for this purpose. Search the Internet for "smartphone binocular adapter."

Left: SnapZoom universal digiscoping adapter for iPhone, Android and Windows smartphones. Compatible with binoculars, spotting scopes, telescopes and microscopes. Right: Carson HookUpz iPhone 4/4S/5/5S or iPhone 6 or iPhone 6 Plus or Samsung Galaxy S4, digiscoping adapter for most full-sized binoculars.

Smartphone cameras are great for capturing the mood of the crowd. Try setting your camera on a tripod or other mount and installing a time lapse photography application. These are available for download from your favorite app store. Set it up to record a snapshot of the crowd every 10 seconds or so and see what you get. Alternatively, if your phone has enough memory, record video of the crowd for about 15 minutes before and during totality. The audio alone will be interesting.

I suggest you pick up a tripod or mount like the GorillaPod for small cameras or smartphones. It will be very handy if you decide to capture video or time-lapse images.

GorillaPod flexible smartphone mount by Joby.

Challenge 2: Exposure

Two different scenarios need to be considered: photographing during the partial phases and photographing totality. Before getting started, set your camera's light meter to examine the entire scene (averaging or smart mode), not spot mode. If you are unsure what this is about, don't worry, this is the default setting on all cameras.

Exposure during the Partial Phases

The Sun is too bright for your camera to capture the partial phases without a special solar filter. Either purchase or make your own filter using material specifically made for solar photography. I do not recommend using the standard neutral density (ND) filters made for everyday photography even if they appear to be dark enough. These will not protect your eyes if your camera has an optical viewfinder.

Camera filters mount in one of three ways: slide-on, clamp-on or screw-on. I find the slide-on and screw-on types to work best.

When purchasing pre-made screw-on camera filters, if you intend to use it on more than one lens, be sure you get the filter that fits directly on the lens with the largest filter diameter. Then purchase an adapter to accommodate your smaller lens.

Just like for your binoculars, you can easily make a camera filter using purposely-made solar safety film readily available on the Internet. To save money, order one larger sheet (perhaps 8 x 10) and use it to make both your camera and binocular filters.

See Chapter 8, "Safely Viewing the Eclipse," for more information. Also see the Chapter 12, "Resources," for places to purchase ready-to-use filters and filters in sheet form.

Taking the Partial Phase Shot

After installing your filter on the front of the camera lens, to get the best image possible with the least amount of noise, choose the lowest

ISO setting on your camera (usually ISO 100). Place the camera in Aperture Priority (AP) mode, setting an aperture between f/2 and f/5.6 and let the camera select the shutter speed. I usually set it to f/5.6.

If your camera has built-in bracketing, turn it on. My Olympus OMD EM-5 lets me take up to seven images at the rate of nine frames per second with each shutter press. If you don't have automatic bracketing, take several shots in succession and adjust your exposure for each shot (by either adding/subtracting via an exposure adjustment setting or by simply changing the shutter speed manually).

If the shutter speed seems too slow (I would consider anything below about 1/100 of a second as too slow), then adjust your aperture accordingly. To test your set-up in advance of the eclipse, go out and shoot the Sun in midday. The brightness of the full Sun is the same as what you'll experience during most of the eclipse's partial phase.

Why can't you shoot without a filter by just selecting the highest shutter speed and setting the darkest aperture? Three answers: First, the intense light can do damage to your camera, or your eyes if you are looking through an optical viewfinder. Second, for many cameras you still won't be able to restrict the light hitting the sensor enough for a good exposure because the Sun is so darn bright. Third, using a very small iris to restrict the light may not provide the sharpest possible image due to the effect of lens diffraction.

Without explaining the math behind diffraction (believe me, you don't want to know), small apertures (large f-stop numbers) from about f11 to f22 will produce less sharp images than the ones below f11. Use a proper solar filter so that you can shoot with a larger aperture (under F11) and avoid losing sharpness to diffraction.

F-Stop: The amount of light reaching your camera's sensor is adjusted by either reducing the amount of time the shutter is open (a faster shutter speed) or by reducing the size of your lens opening. Reducing the lens opening is accomplished by making the "iris" of the lens smaller (reducing the aperture). If you reduce the aperture by one full f-stop, you are letting in half as much light as you were at the prior setting. Note that most cameras have settings in increments of one-third or one-half f-stop. The full f-stop settings are as follows: f/1.4, f/2, f/2.8, f/4, f/5.6, f/8, f/11, f/16, f/22. The in-between settings on your camera are just that, they sit between the full f-stops in brightness.

The higher f-stop number means a smaller opening and, therefore, less light. This is because the f-stop number is a fraction. Because what we call f/2 and f/8 are really 1/2 and 1/8 respectively, the latter f/8 is a smaller opening and lets in less light. The bigger the f-number, the less light gets to the sensor.

When shooting Baily beads and the diamond ring, be sure to use a proper solar filter. If you have set up automatic bracketing, you should get great results. With manual bracketing, plan on adjusting your exposure to let in more light as the Moon covers more of the entire Sun. Remember to not disturb your focus setting when adjusting the aperture.

Exposure during Totality

During totality, you need to be flexible. The dynamic range presented by the totally eclipsed sky is much broader compared to what your camera is capable of capturing in a single exposure (dark moon and bright corona). This means you'll need to take lots of pictures and possibly combine them together later on your PC to get something representative of what you witnessed in person. For example, for the corona photo, will the surface of the moon appear as a black disk or will you pick up the fine details on the Moon's surface using a longer exposure?

Also, the inner portion of the corona will present a much brighter image than the outer edges. If you want to capture its fine details, be prepared to bracket.

A tripod is essential for good images of totality. With a tripod, it's easier to keep the Sun centered in frame when using a large focal-length lens. It will also help you to maintain sharp focus from shot to shot.

> Be nice to your fellow eclipse observers: Totality may be the first time your camera has seen dark conditions all day. Make sure your flash is disconnected from the camera or it is positively turned off.

Taking the Totality (Corona) Shot

Just after totality starts, remove your filter and take a moment and re-establish your exposure settings. The good news is that the light level will remain stable throughout totality.

Start with the lowest ISO your camera will support (ISO 100) to minimize noise. In aperture priority (AP) mode, select the widest open aperture of your lens. Let the camera choose an appropriate shutter speed and then if the shutter speed seems too low (under 1/100), bump up the ISO setting until the camera is "happy" with the exposure. Try to keep ISO under 400. With my camera, I can go as high as ISO 650 and still get excellent results. If all you want is a snapshot, even higher ISOs will be fine.

The question that may arise at this point is: What is too low a shutter speed? After all, your camera is mounted on a tripod, so why not just make a 10-second exposure and keep the ISO down at 100? The answer is that the Earth, the Moon and the Sun are all moving.

At very slow shutter speeds, the motion of the eclipse objects is going to soften your picture. If time permits during totality, try taking low ISO pictures at shutter speeds slower than one second and see how they come out.

Once you've established your nominal exposure settings, use the automatic bracketing feature of your camera to capture the dynamic range of the scene. In aperture priority mode (AP), auto bracketing will vary the shutter speed, producing different exposures. Without auto-bracketing, you should still take several sets of shots around a handful of manually adjusted shutter speeds and aperture settings.

You should vary shutter speed and aperture not to get a "correctly exposed" image, but to capture different images of the same scene. You'll get interesting photos at all exposure settings. During totality, shorter exposure times will show solar prominences and longer ones will reveal the full extent of the corona.

If you don't have auto-bracketing, use your manual exposure compensation dial or buttons. Review your results on the back of the camera and be sure you have at least one of the following images:

- A shot bright enough to see the surface features of the moon.
- A shot dark enough to see the edges of the corona with a dark sky and good resolution between the corona's filaments.
- A shot showing the solar prominences.

HDR (High Dynamic Range) and Auto-bracketing: Consider the following scene: A picture to be taken from the inside of a barn (dark inside) with a horse grazing in a field visible through the barn door (bright outside). The scene to be photographed has a broad range of bright and dark shadow areas that are too extreme for a camera to capture in one shot. Cameras have exposure adjustments so we can decide which of these extreme areas we just won't try to capture in the particular shot we are about to take. While the dynamic range of cameras continuously improve, it has not yet matched the capabilities of human sight.

HDR photography attempts to address this shortcoming by combining shots of the same scene that are taken with various exposure settings. Some shots capture the brightest parts of the scene, and some capture the darkest parts. Afterward, these shots (as few as three and as many as seven) are combined into a single image showing the full dynamic range in the original scene.

Older cameras don't have an HDR feature, and newer cameras that support HDR deal with it in one of two ways: Some have an HDR setting that tells the camera to take a handful of images in rapid succession with varying exposures. Then the camera assembles these into a single HDR image. The second approach uses the auto-bracketing feature of the camera to capture the images and then leaves it to the photographer to assemble them into a single image using a computer.

For HDR photography, set the auto-bracketing feature for a constant aperture and let the camera automatically vary shutter speed to accomplish the required range of exposures. This helps ensure a sharp final composite image because the depth-of-field (focus depth) is consistent for all the shots.

The "auto-bracket" feature was added to cameras well before the advent of HDR photography. It is used by photographers to get "insurance" shots in case their exposure was misadjusted.

HDR images can be made to look either natural or artistic. Artistic HDR images are produced in PC software using a process called "tone mapping." Early HDR examples made heavy use of tone mapping, resulting in unnatural looking images. When looking at an HDR image processed to look natural, you may not notice that it's an HDR.

Challenge 3: Focus

Tack Sharp Photos

To get tack sharp images, use a good tripod. A good one will stand up to moderate winds and will generally use a ball type head with a smooth action so you can quickly make adjustments without mechanical slippage getting in the way.

The second item I highly suggest using is a remote shutter trigger. This allows you to snap the picture without actually touching the camera, eliminating vibrations. Your camera manufacturer is the best place to get one. They come in two different types; wireless or wired. Either one will do the job. If you will be using your camera's auto-bracket function (which I highly recommend) be sure and test how your remote trigger works when the camera is set to auto-bracket. You may need to hold the remote trigger down until the last shot has been taken.

If you don't have a remote trigger, the next best thing is to use your camera's self-timer. Set up the camera so that when you press the shutter release, the camera counts down and gives time for vibrations to settle out before the picture is taken. Most cameras have two different timer lengths, one is long enough for you to jump into the picture, wrap your arm around the person next to you and flash a great big smile (about ten seconds). The other is much shorter and is designed specifically for this anti-vibration application (usually two seconds). With many cameras, the self-timer can't be used with auto-bracketing.

During the total eclipse, temperatures will drop 10 to 20 degrees. This temperature swing is sufficient to change the mechanical components of your camera-lens combination, throwing it off focus. Telescope operators have the same problem. The solution is to check and readjust focus frequently throughout the event.

Auto Focus or Manual Focus

Eclipse photographers will tell you it is best to focus manually. One of three strategies are employed:

1) Manual focus on the Moon-Sun.

2) Manual focus on infinity.

3) Auto focus on the Moon-Sun then switch to manual focus and adjust as necessary.

If you've never used the manual focus on your digital camera, make sure you practice before eclipse day. Shoot until you are comfortable with the procedure, verifying your images to make sure they are sharp.

Some cameras have manual focus assist modes. The Olympus OMD EM-5 has the ability to enlarge the image in the viewfinder and on the rear display so you can easily verify focus in manual mode. It pays to crack open your camera manual to see if there are any features designed to make manual focus easier.

Whatever method you choose to use, the most important procedural step to remember is to look at an enlarged image of your picture on the camera's display and verify focus. If you don't know how to zoom the image on your rear display, learn and practice the technique.

Video

Shooting video is a great way to capture the mood of the crowd during and immediately after the eclipse. Set up a video camera and I assure you the audio alone will be well worth the effort. This might be the best use of your smartphone camera. Just make sure you have enough digital memory and budget your capture time accordingly.

You can shoot video of the Sun and Moon, but remember, you will need an appropriate solar safety filter. Some camcorders have powerful zoom lenses that will render a decent size image. Instead of a camcorder, you can use a digital camera with video recording capability. A video camera with a high focal length lens can be especially effective. Set focus manually or the camera may end up hunting for a lock at the most important moments (when the light changes abruptly). Don't forget to remove the filter during totality.

Conclusion

At the risk of being repetitive, I will wrap-up this chapter on photographing the eclipse with one final thought. Landscape photographers believe that the best time for taking a picture is during the "magic hours"; the first hour after sunrise and the first hour before sunset. The light during these times are warm in color, billowy clouds often appear, and shadows are long and soft. This gives us two opportunities every day of the year to take advantage of ideal outdoor natural lighting conditions.

Similarly, the first two minutes after totality ends will be "magic minutes" for people photographers. Everyone's face will be lit up in pure joy and amazement. You will be surrounded by big natural smiles, fully open eyes, and spontaneous hugging. Some of the most cherished photographs (and videos clips) in my collection were taken during those precious few minutes. Be prepared to capture this with your camera. Unlike the landscape photographer's "magic hours," this opportunity won't be happening twice each day!

In addition to the awe and amazement you will feel during the eclipse, you may discover—as I did—an unexpected powerful urge to share the experience with others. This book was inspired by that urge.

I hope the information and guidance provided in "Total Solar Eclipse 2017: The Next US Eclipse," has been interesting and proves useful on your adventure. I would be forever grateful if you would take a moment and leave a review of this book on Amazon at http://amazon.com/author/marcnussbaum.

Happy eclipse chasing!

Chapter 12. Resources

Selected websites maintained by the eclipse community

1) GreatAmericanEclipse.com is published by Michael Zeiler and Polly White to educate the public on how to witness nature's greatest spectacle, a total eclipse of the Sun:

> *http://www.GreatAmericanEclipse.com*

2) Eclipse2017.org was established to provide information to news media, elected officials, school administrators and teachers, and the general public as to exactly how to view the magnificent total eclipse of the sun in 2017. The site is maintained by Dan McGlaun (a veteran of eleven total solar eclipses):

> *http://www.eclipse2017.org*

3) NASA's (National Aeronautics and Space Administration) official site for the 2017 eclipse. Eclipse shadow maps, local circumstances tables and more:

> *http://eclipse.gsfc.nasa.gov/SEmono/TSE2017/TSE2017.html*

4) NASA's main website has fantastic images, including a live stream of video from the International Space Station. Here is where to go for the latest on the US space program:

> *http://www.nasa.gov*

5) NASA Goddard Space Center YouTube page. Excellent educational videos:

> *https://www.youtube.com/channel/UCAY-SMFNfynqz1bdoaV8BeQ*

6) Fred Espenak's latest eclipse websites:

> *http://www.mreclipse.com*
>
> *http://eclipsewise.com*

7) Glenn Schneider's "umbraphile" eclipse page:

http://nicmosis.as.arizona.edu:8000/UMBRAPHILLIA.html

Links for local circumstances timetables

1) Click on a Google map location to get local eclipse timetables:

http://eclipse.gsfc.nasa.gov/SEgoogle/SEgoogle2001/SE2017Aug21Tgoogle.html

2) Nice table driven interface by state:

http://www.eclipse2017.org/2017/local_circumstances.htm

3) Eclipse-Chasers.com is written and maintained by Bill Kramer as a public service. The webpage has a calculator for local circumstances that takes into account the lunar limb profile:

https://www.eclipse-chasers.com/php/tseCalculator.php

Eye protection suppliers

1) Rainbow Symphony, Inc., 6860 Canby Ave. #120, Reseda, CA 91335 (818) 708-8400:

http://www.rainbowsymphonystore.com/eclipseshades.html

2) Eclipse2017.org:

http://www.eclipse2017.org/glasses_order.htm

3) Thousand Oaks Optical, Kingman Arizona:

http://www.thousandoaksoptical.com/ecplise.html

4) Baader Planetarium AstroSolar at Adorama. Also available on Amazon:

http://www.adorama.com/AA2459283.html?gclid=CKeg_aKN1MYCFYgBaQodijUKbQ

Equipment suppliers

1) B and H online store for cameras, binoculars, etc. (located in NYC). Here's the link to their excellent and informative guide on how to buy a pair of binoculars:

   http://www.bhphotovideo.com/explora/photography/buying-guide/binocular

2) Celestron International, 2835 Columbia St., Torrance, CA 90503. (310) 328-9560:

   http://www.celestron.com

3) Meade Instruments Corporation, 16542 Millikan Ave., Irvine, CA 92714. (714) 756-2291:

   http://www.meade.com

4) Orion Telescopes and Binoculars, 89 Hangar Way, Watsonville, CA 95076 (831) 763-7000:

   http://www.telescope.com/home.jsp

Binoculars and camera filter supplier websites (pre-made filters and sheets)

These items are also available through Amazon and other on-line resellers.

1) Bader Planetarium, German manufacturer of filters and safety film:

   http://www.baader-planetarium.com

2) AstroSolar filter manufacturer and materials:

   http://astrosolar.com/en

   http://astrosolar.com/en/information/about-astrosolar-solar-film/differences-in-astrosolar-solar-films

3) Kendrick Astronomy:

http://www.kendrickastro.com/solarfilters.html

4) Thousand Oaks Optical, Box 4813, Thousand Oaks, California 91359 (805) 491-3642:

http://www.thousandoaksoptical.com/ecplise.html

Binocular review and buying guides

1) Nikon has an excellent website explaining everything you want to know about binoculars:

http://www.nikon.com/products/sportoptics/how_to/guide/binoculars/basic/basic_07.htm

2) This website explains angle of view specifications for binoculars:

http://www.bestbinocularsreviews.com/wide-angle-binoculars.php

3) Article by Dave Brody at space.com on "How to Choose Binoculars for Astronomy and Sky watching":

http://www.space.com/27404-binoculars-buying-guide.html

4) Online calculator to convert binocular field of view between feet (meters) and angular units:

http://astronomy.tools/calculators/binoculars

Instructions for making camera and binocular filters

1) Baader Planetarium's "How to make your own objective solar filter for your camera, telescope, spotting scope or binocular":

http://astrosolar.com/en/information/how-to/how-to-make-your-own-objective-solar-filter-for-your-camera-or-telescope

2) Jason McCabe's YouTube video "How to build a Solar Filter for DSLR" cameras:

https://www.youtube.com/watch?v=2uFlOCyPBlU

Weather Information websites

1) Jay Anderson's eclipse weather website:

   http://www.eclipser.ca

2) U.S. National Weather Service:

   http://www.weather.gov

3) National Oceanic and Atmospheric Administration (NOAA):

   http://www.noaa.gov

4) National Center for Environmental Prediction (part of NOAA):

   http://www.ncep.noaa.gov

Organized eclipse tours

1) Sky and Telescope:

   http://www.skyandtelescope.com/astronomy-travel/sky-telescope-2016-2017-eclipse-tours

2) Travelquest:

   http://www.travelquesttours.com/tours/2017-national-parks-of-the-american-west-total-solar-eclipse/welcome

3) Search the Web for "2017 eclipse tour."

Appendix: Maps

U.S. Overview

Eclipse path through the United States. Maximum totality is 2 minutes, 40 seconds near Carbondale, Illinois. Observers inside the narrowly spaced red lines will see a total eclipse. Between the red lines and the first black lines the eclipse will be partial with between 99% and 80% of the Sun covered. People within the next band of black lines will see between 80% and 60% of the Sun covered.

Pacific and Mountain States Region

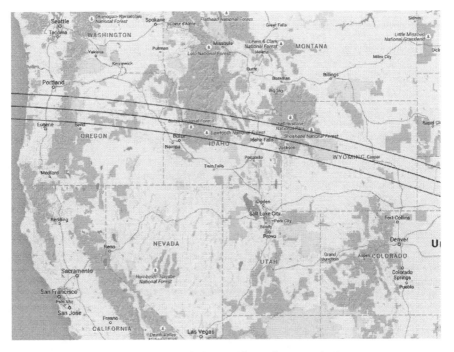

Eclipse path through the Pacific and Mountain States.

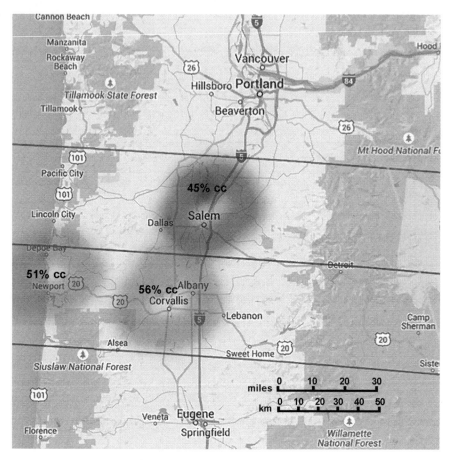

Eclipse path through Western Oregon.
Cloud cover key: red= greater than 50%, blue= 40-50%, green= less than 40.

Oregon 2 of 3

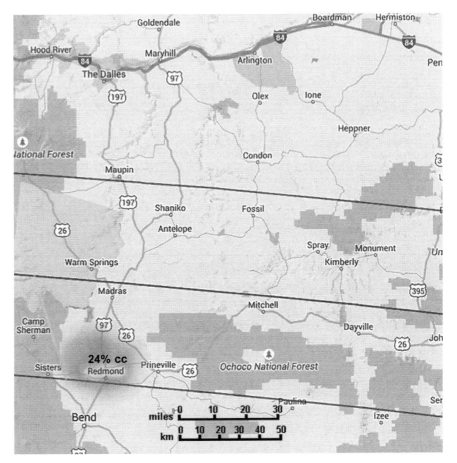

Eclipse path through Central Oregon.
Cloud cover key: red= greater than 50%, blue= 40-50%, green= less than 40%.

Oregon 3 of 3

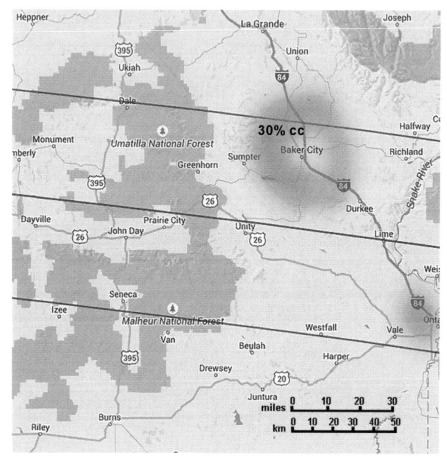

Eclipse path through Eastern Oregon.
Cloud cover key: red= greater than 50%, blue= 40-50%, green= less than 40%.

Idaho 1

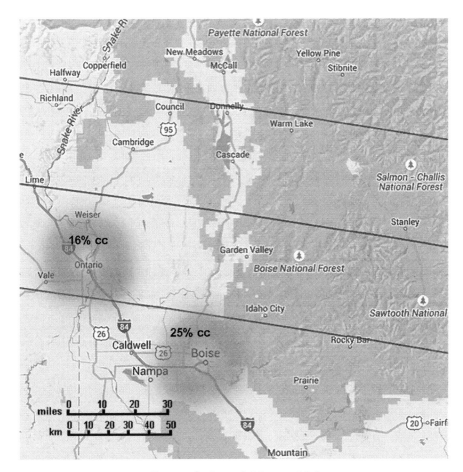

Eclipse path through Western Idaho.
Cloud cover key: red= greater than 50%, blue= 40-50%, green= less than 40%.

Idaho 2

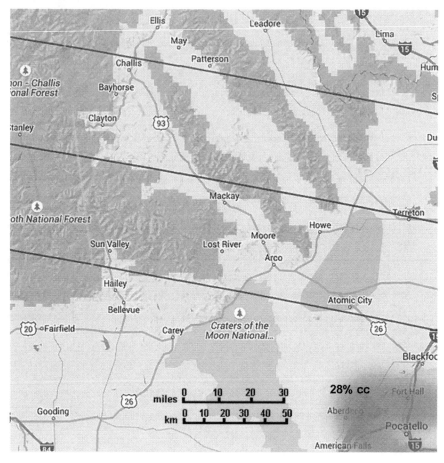

Eclipse path through Central Idaho.
Cloud cover key: red= greater than 50%, blue= 40-50%, green= less than 40%.

Idaho 3

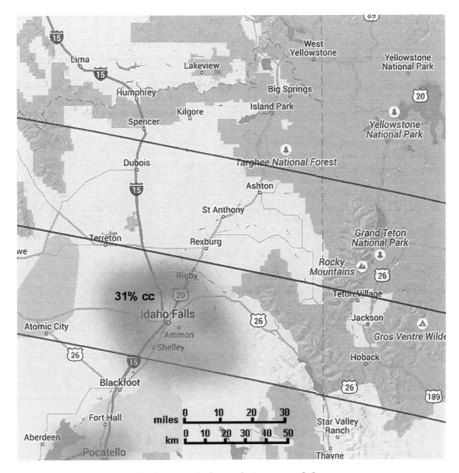

Eclipse path through Eastern Idaho.
Cloud cover key: red= greater than 50%, blue= 40-50%, green= less than 40%.

Wyoming 1

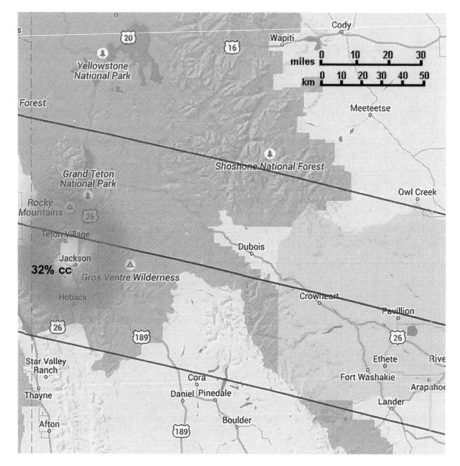

Eclipse path through Western Wyoming.
Cloud cover key: red= greater than 50%, blue= 40-50%, green= less than 40%.

Wyoming 2

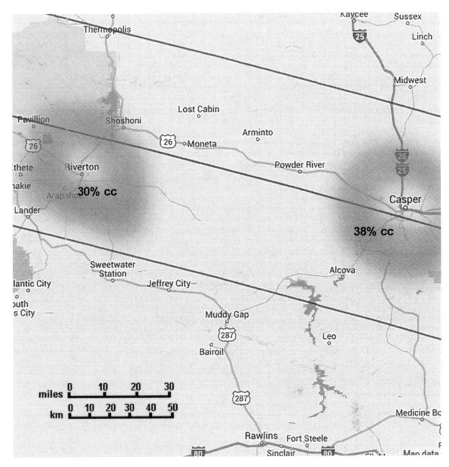

Eclipse path through Central Wyoming.
Cloud cover key: red= greater than 50%, blue= 40-50%, green= less than 40%.

Wyoming 3

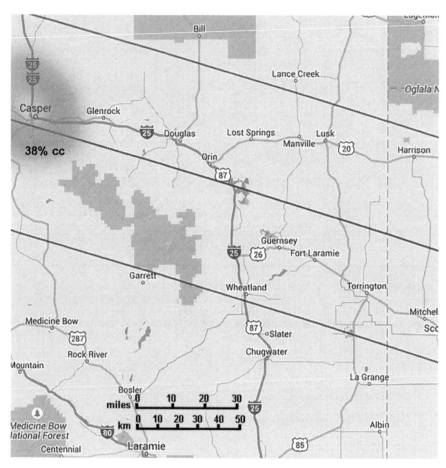

Eclipse path through Eastern Wyoming.
Cloud cover key: red= greater than 50%, blue= 40-50%, green= less than 40%.

Central States Region

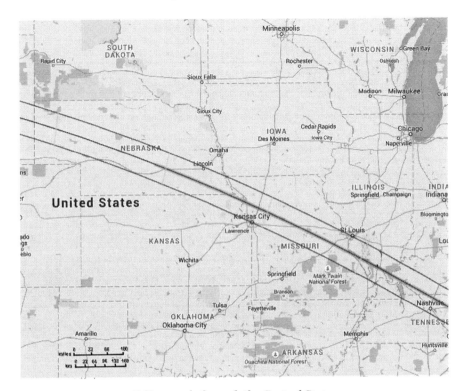

Eclipse path through the Central States.

Nebraska 1

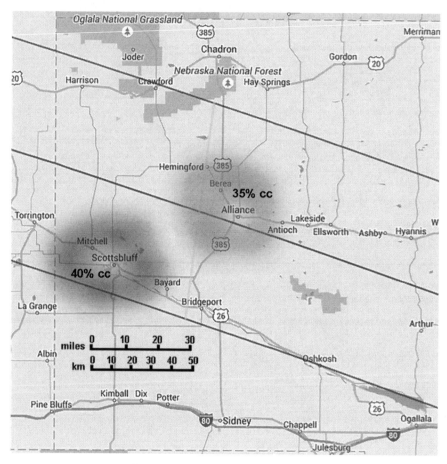

Eclipse path through Western Nebraska.
Cloud cover key: red= greater than 50%, blue= 40-50%, green= less than 40%.

Nebraska 2

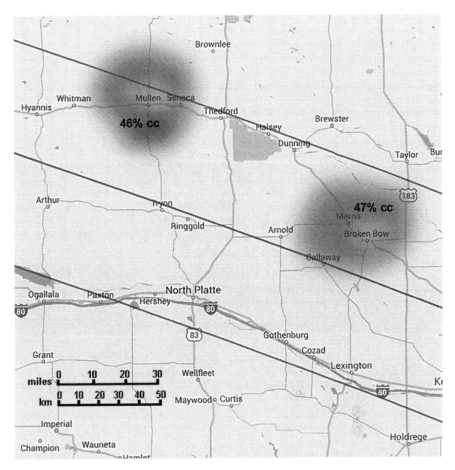

Eclipse path through Central Nebraska.
Cloud cover key: red= greater than 50%, blue= 40-50%, green= less than 40%.

Nebraska 3

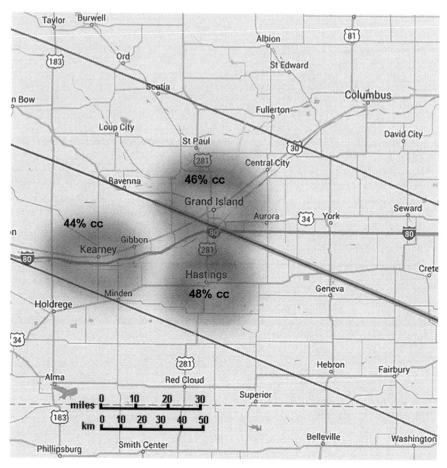

Eclipse path through Central Nebraska.
Cloud cover key: red= greater than 50%, blue= 40-50%, green= less than 40%.

Nebraska 4

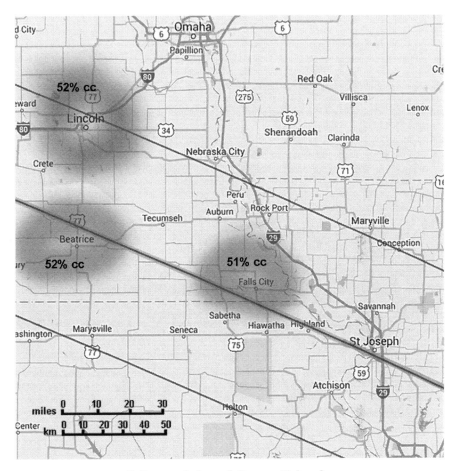

Eclipse path through Eastern Nebraska.
Cloud cover key: red= greater than 50%, blue= 40-50%, green= less than 40%.

Missouri 1

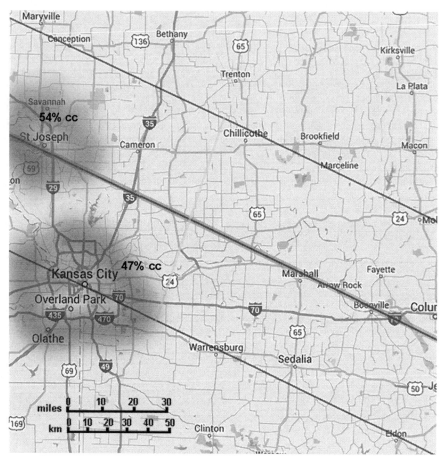

Eclipse path through Western Missouri.
Cloud cover key: red= greater than 50%, blue= 40-50%, green= less than 40%.

Missouri 2

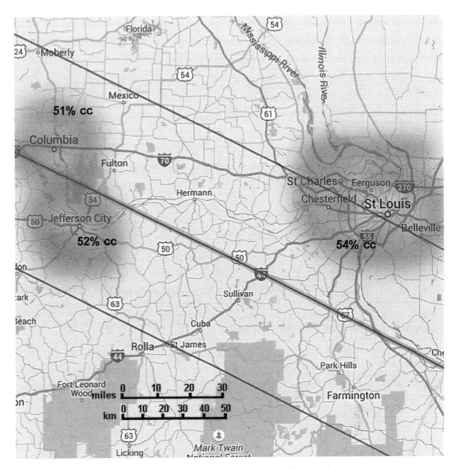

Eclipse path through Eastern Missouri.
Cloud cover key: red= greater than 50%, blue= 40-50%, green= less than 40%.

Illinois 1

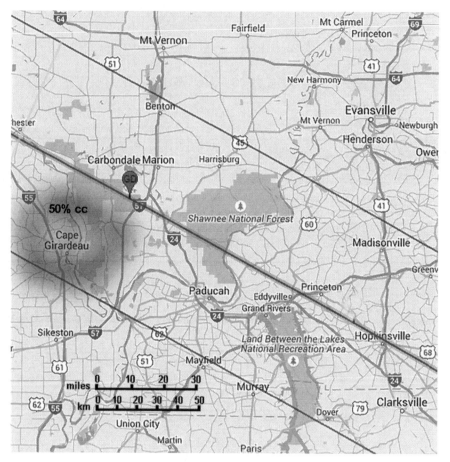

Eclipse path through Illinois. Carbondale, Illinois will experience greatest eclipse duration "GD" of 2 minutes, 40 seconds.
Cloud cover key: red= greater than 50%, blue= 40-50%, green= less than 40%.

Eastern States Region

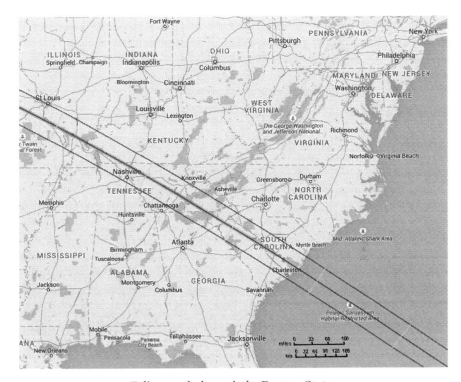

Eclipse path through the Eastern States.

Tennessee 1

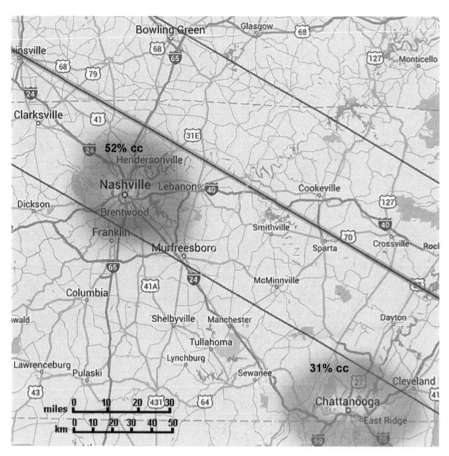

Eclipse path through Western Tennessee.
Cloud cover key: red= greater than 50%, blue= 40-50%, green= less than 40%.

Tennessee 2

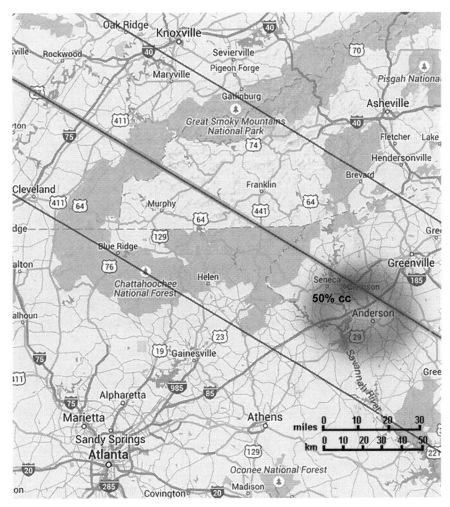

Eclipse path through Eastern Tennessee.
Cloud cover key: red= greater than 50%, blue= 40-50%, green= less than 40%.

South Carolina 1

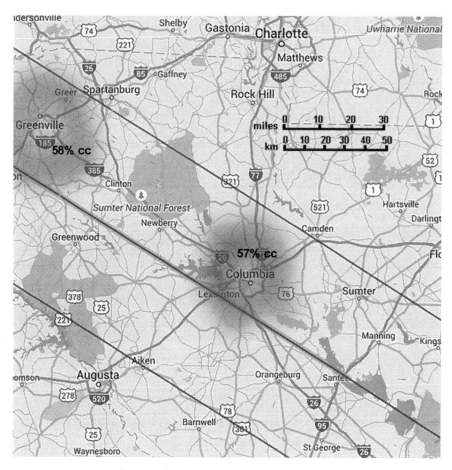

Eclipse path through Western South Carolina.
Cloud cover key: red= greater than 50%, blue= 40-50%, green= less than 40%.

South Carolina 2

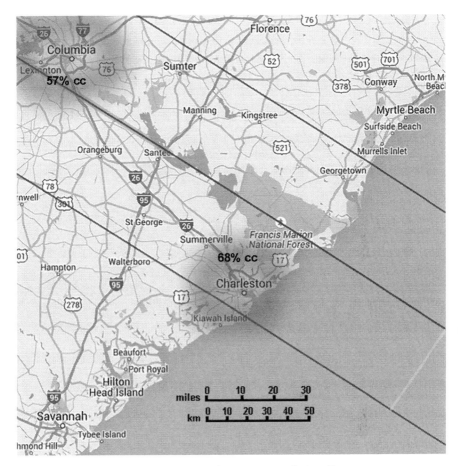

Eclipse path through Eastern South Carolina.
Cloud cover key: red= greater than 50%, blue= 40-50%, green= less than 40%.

Endnotes

[1] Jessica Orwig "Why We Will Never Build a Nuclear Fusion Reactor As Good As the Sun," Business Insider, October 16, 2014, accessed June 8, 2015, < http://www.businessinsider.com/we-will-never-have-sun-like-nuclear-fusion-2014-10>

[2] Fred Espenak, "Five Millennium Catalog of Solar Eclipses," 2001 - 2100), NASA, <http://eclipse.gsfc.nasa.gov/SEcat5/SEcatalog.html>

[3] "Destination Star Trek," creator of Star Trek, Gene Roddenberry, was given this nick name by original series associate producer Robert Justman, accessed June 22, 2015,   <http://www.destinationstartrek.com/ten-forward/news/495-the-great-bird-of-the-galaxy>

[4] Douglas Adams, "Last Chance to See"

[5] "Last Chance to See," Wikipedia encyclopedia, accessed June 11, 2015, <https://en.wikipedia.org/wiki/Last_Chance_to_See>

[6] "Actuarial Life Table for 2010," Social Security Website, accessed June 12, 2015, <http://www.ssa.gov/oact/STATS/table4c6.html>

[7] "Bartender Terminology," Wikipedia encyclopedia, accessed June 8, 2015, <https://en.wikipedia.org/wiki/Bartending_terminology>

[8] "FAQ" Eclipse2017.org, accessed June 8,2015, <http://www.eclipse2017.org/2017/faq.htm>

[9] "Sun," NASA, accessed June 29, 2015 <http://www.nasa.gov/sun>

[10] "Terms and Acronyms," NASA, <http://www.nasa.gov/mission_pages/iris/overview/definitions_prt.htm>

[11] "Future of the Earth," Wikipedia encyclopedia, accessed June 29, 2015, <https://en.wikipedia.org/wiki/Future_of_the_Earth>

[12] "Northern Lights," The Northern Lights Center, accessed July 8, 2015, <http://www.northernlightscentre.ca/northernlights.html>

[13] Monthly Notices of the Royal Astronomical Society, Volume IV, December 9, 1836, Accessed on June 11, 2015, <http://mnras.oxfordjournals.org/>

[14] "How is the Corona Heated?", accessed July 16,2015, <http://hesperia.gsfc.nasa.gov/sftheory/heat.htm>

[15] "SOHO fact sheet" http://www.nasa.gov/pdf/156578main_SOHO_Fact_Sheet.pdf

[16] "Solar Probe Plus," NASA Goddard Space Flight Center, accessed July 16,2015, <http://solarprobe.gsfc.nasa.gov/spp_mission.htm>

[17] Merriam-Webster online dictionary accessed May 21, 2015, search term: eclipse, <http://www.merriam-webster.com>

[18] Merriam Webster online dictionary accessed May 23, 2015, search term: syzygy, <http://www.merriam-webster.com>

[19] "Star Trek II: The Wrath of Khan," motion picture, 1982

[20] "Ecliptic," Dictionary.com, accessed July 7, 2015,
<http://dictionary.reference.com/browse/ecliptic>

[21] John Walker, "Moon near Perigee, Earth near Aphelion," Fourmilab, Retrieved June 11, 2015, <http://www.fourmilab.ch/images/peri_apo>

[22] "Total Eclipses," Ayman Mohamed Ibrahem, retrieved April 2015,
<http://www.bibalex.org/eclipse2006/TotalEclipses.htm>

[23] Guillermo Gonzales, "Mutual Eclipses in the Solar System," News and Reviews in Astronomy and Geophysics, Oxford Journals.
<http://astrogeo.oxfordjournals.org/content/50/2/2.17.full>

[24] "Voyage to Darkness– In Search of Eclipses,"
<http://www.nauticom.net/www/planet/files/EclipseHistory-AstronomyThemeCruises.html> accessed June 27, 2015

[25] Jay Anderson, http://www.eclipser.ca

[26] Tom Jones, "Never Say No," Lyrics from the Fantasticks, The Broadway Musicals.com, accessed July 9, 2015, <
http://www.thebroadwaymusicals.com/lyrics/fantasticksthe/neversayno.htm>

[27] William Shakespeare, "Othello," Act 5 scene 2.

[28] Bonnie Tyler and Jim Steinman, "Total Eclipse of the Heart," Wikipedia encyclopedia, accessed June 28,2015,
<https://en.wikipedia.org/wiki/Total_Eclipse_of_the_Heart>

[29] Carly Simon, "You're So Vain," Wikipedia encyclopedia, accessed June 28, 2015, <https://en.wikipedia.org/wiki/You%27re_So_Vain> and "Boys in the Trees," published November 24, 2015, Flatiron Books.

[30] Nicholas Clee, "Eclipse: The Horse That Changed Racing History Forever," The Overlook Press; 1 edition (March 29, 2012)

[31] "Solar Eclipse of October 27, 1780, Wikipedia encyclopedia, accessed June 28, 2015, <https://en.wikipedia.org/wiki/Solar_eclipse_of_October_27,_1780>

[32] Dyson and Eddington, 1920. "A Determination of the Deflection of Light by the Sun's Gravitational Field." From observations made at the total eclipse of 29 May, 1919. Royal Society. Philosophical Transactions.

[33] NASA, "Did ancient peoples really predict solar eclipses?", accessed August 4, 2015, <http://image.gsfc.nasa.gov/poetry/ask/a11846.html>

[34] Fred Espenak, "Five Millennium Catalog of Solar Eclipses, 2001 - 2100"), NASA, <http://eclipse.gsfc.nasa.gov/SEcat5/SEcatalog.html>

[35] Fred Espenak, "Eclipses and the Saros," (August 28, 2009), NASA Goddard Space Flight Center, <http://eclipse.gsfc.nasa.gov/SEsaros/SEsaros.html>

Made in the USA
Columbia, SC
26 April 2017